Environment and Food

This timely book provides a thorough introduction to the interrelationship of food and the environment. Its primary purpose is to bring to our attention the multiplicity of links and interconnections between what we eat and how this affects the Earth's resources. Having a better idea of the consequences of our food choices might encourage us to develop more sustainable practices of production and consumption in the decades ahead.

Although human societies have, over time, brought under control a large proportion of the Earth's resources for the purpose of food production, we remain subject to the effective functioning of global ecosystem services. The author highlights the vital importance of these services, and explains why we should be concerned about the depletion of freshwater resources, soil fertility decline and loss of biological diversity. The book also tackles some of the enormous challenges of our era: climate change, to which the agri-food system is both a major contributor and a vulnerable sector; and the prospect of significantly higher energy prices, arising from the peaking of oil and gas supplies, which will reveal how dependent the food system has become upon cheap fossil fuels. Such challenges are likely to have significant implications for the long-term functioning of global supply chains, and raise profound questions regarding the nutritional security of the world's population. This book argues that a re-examination of the assumptions and practices underpinning the contemporary food system is urgently required.

Environment and Food is a highly original, interdisciplinary and accessible text that will be of interest to students and the wider public genuinely interested in, and concerned by, the state of the world's food provisioning system. It is richly illustrated with figures and makes extensive use of boxes to highlight relevant examples.

Colin Sage is Senior Lecturer in Geography at University College Cork, Republic of Ireland.

Routledge Introductions to Environment Series
Published and forthcoming titles

Environmental Science texts

Atmospheric Processes and Systems
Natural Environmental Change
Environmental Biology
Using Statistics to Understand the
  Environment
Environmental Physics
Environmental Chemistry
Biodiversity and Conservation,
  2nd Edition
Ecosystems, 2nd Edition
Coastal Systems, 2nd Edition

Titles under Series Editor:
David Pepper

Environment and Society texts

Environment and Philosophy
Energy, Society and Environment,
  2nd Edition
Gender and Environment
Environment and Business
Environment and Law
Environment and Society
Environmental Policy
Representing the Environment
Sustainable Development
Environment and Social Theory,
  2nd Edition
Environmental Values
Environment and Politics, 3rd Edition
Environment and Tourism, 2nd Edition
Environment and the City
Environment, Media and
  Communication
Environmental Policy, 2nd Edition
Environment and Economy
Environment and Food

Forthcoming

Environmental Governance

Environment and Food

Colin Sage

 Routledge
Taylor & Francis Group

LONDON AND NEW YORK

First published 2012
by Routledge
2 Park Square, Milton Park, Abingdon, Oxon OX14 4RN

Simultaneously published in the USA and Canada
by Routledge
711 Third Avenue, New York, NY 10017

Routledge is an imprint of the Taylor & Francis Group, an informa business

British Library Cataloguing in Publication Data
A catalogue record for this book is available from the British Library

Library of Congress Cataloging in Publication Data
Sage, Colin.
  Environment and food/Colin Sage.
    p. cm.
    1. Food habits – Environmental aspects.   2. Food preferences –
  Environmental aspects.   3. Food supply – Environmental aspects.
    I. Title.
  GT2850.S24 2011
  363.8′72 – dc22   2010052901

ISBN: 978–0-415–36311–2 (hbk)
ISBN: 978–0-415–36312–9 (pbk)
ISBN: 978–0-203–01346–5 (ebk)

Typeset in Times New Roman
by Florence Production Ltd, Stoodleigh, Devon

Printed in the United Kingdom
by Henry Ling Limited

Contents

List of plates   ix

List of figures   x

List of tables   xi

List of boxes   xii

Acknowledgements   xiv

Chapter 1   Introduction: why environment and food?   **1**
Food and the environment   *3*   ● Sustainability and
food   *4*   ● Focus and structure of the book   *7*
● Further reading   *12*

Chapter 2   The global agri-food system   **14**
Development   *16*   ● Scale and structure   *27*
● Primary food production   *32*   ● Agri-technologies   *41*
● Food manufacturing   *49*   ● Food retail   *54*
● Summary   *64*   ● Further reading   *65*

Chapter 3   The agro-ecology of primary food production   **67**
Ecosystem services   *69*   ● Typology of agricultural
systems   *74*   ● Resources   *90*   ● Summary   *108*
● Further reading   *109*

Chapter 4   Global challenges for food production   **111**
Climate change   *112*   ● Freshwater   *120*   ● Peak
oil   *132*   ● Livestock   *141*   ● Summary   *153*
● Further reading   *154*

Chapter 5   Final foods and their consequences   **156**
Transforming foods   *158*   ● Environmental dimensions   *167*
● Transporting food   *187*   ● Food waste   *199*
● Summary   *207*   ● Further reading   *208*

Chapter 6   Rethinking food security   **209**
    Evolution   *211*   ●   Population   *218*   ●   Food and energy
security   *222*   ●   Climate change   *226*   ●   Globalisation   *233*
  ●   Food sovereignty   *241*   ●   Summary   *246*   ●   Further
reading   *247*

Chapter 7   Towards a sustainable agri-food system   **249**
    Reconsidering sustainability   *251*   ●   Sustainable agriculture   *255*
  ●   Sustainable consumption   *262*   ●   Reconnecting production
and consumption   *271*   ●   Planning food systems   *277*
  ●   Summary   *290*   ●   Further reading   *291*

Chapter 8   Conclusion   **293**

Glossary   297
Bibliography   300
Index   317

Note: Words highlighted on their first occurrence in the text can be found in the glossary after chapter eight.

Plates

3.1 Aspects of wet rice cultivation, West Sumatra, Indonesia   81
3.2 Pre-Hispanic terracing, Colca Valley, Peru   85
3.3 Agro-forestry: harvesting fruits, Lampung, Indonesia   87
3.4 Ducks as part of integrated pest management, West Sumatra   89
3.5 Aquaculture: cultivating fish in irrigation canals, West Sumatra 91

Figures

2.1	The global agri-food system: a schematic representation	31
2.2	The farm as hub of the agri-food system	38
2.3	Stages in the process of motomechanisation in grain cultivation	43
2.4	The hourglass model of the agri-food system: retailer power	58
3.1	A framework for evaluating primary food production options	78
3.2	A conceptual water-accounting framework	99
4.1	Greenhouse gas emissions from the UK food chain, 2006	114
4.2	Sources of agricultural greenhouse gas emissions	118
4.3	Share of global field-level water use by different crops	128
4.4	Virtual water flows: leading exporters and importers	130
4.5	Discovery and production of regular oil	136
4.6	The depletion curve for oil and gas	138
4.7	World annual fuel ethanol production, 1975–2009	139
5.1	The "co-evolutionary" development of social, economic and technological change	162
5.2	Flowchart of life-cycle phases for a cottage pie ready-meal	186
5.3	Flowchart for Swedish tomato ketchup	188
5.4	CO_2 emissions by mode of transport, UK 2006	192
5.5	Changes in control over the food distribution system, UK	193
6.1	World grain production per person, 1950–2007	219
6.2	World grain stocks as days of consumption, 1960–2007	223

Tables

2.1	Global food sales, 2002	28
2.2	IGD grocery turnover league, 2007	59
3.1	Production statistics for five food and feed commodities, 2007	75
3.2	A global typology of primary food production systems	80
3.3	Energy inputs in US maize (corn) production	107
4.1	Global average virtual water content of selected primary foods	127
4.2	Meat consumption per capita by region	143
4.3	Population of the principal commercial livestock categories by world region	146
5.1	Environmental impacts related to the final consumption of products	184
5.2	Food transport indicators, UK 1992–2006	191
5.3	Illustrations of waste in the food chain	200

Boxes

2.1	Sweetness and sugar	17
2.2	King Sugar and other agricultural monarchs	18
2.3	Observations on the global dynamics of the agri-food system	26
2.4	The European agricultural revolution	33
2.5	Traditional low-input farming, pre-war UK	36
2.6	Features of the agricultural treadmill	39
2.7	The Green Revolution imperative	40
2.8	Stages in the process of motomechanisation	42
2.9	Concentration ratios for selected US food sectors	52
2.10	Some key features of the leading retailers	61
3.1	Millennium Ecosystem Assessment: main findings	70
3.2	Deforestation in Amazonia and soybean production	73
3.3	Total land requirements	74
3.4	Bolivia: potatoes and maize	101
3.5	Energy analysis	105
3.6	A comparison of energy inputs to land preparation	106
4.1	Carbon sequestration options	119
4.2	Mining the Ogallala aquifer	122
4.3	Damming the rivers	123
4.4	Draining the Aral Sea	124
4.5	Canada's tar sands: dirty oil	137
4.6	Bovine culture	145
4.7	Feedlot beef	148
4.8	A pig's life	150
5.1	The fractionation of corn	164
5.2	The development of the modern white sliced loaf	165
5.3	Water use	169
5.4	The global warming potential of refrigerants	173
5.5	The environmental impacts of a cola can	177
5.6	Bottled water: bad for people and the environment?	180
5.7	Evaluating the relative environmental effects of glass bottle re-use	181
5.8	Food air freight: chill chain via plane	195
5.9	Ecologies of scale in transport	198
5.10	Waste carrots	202
5.11	Luxus consumption	206

6.1	Understanding people's entitlement to food	212
6.2	Food poverty	216
6.3	Economic growth and dietary change	220
6.4	The world's growing food-price crisis	223
6.5	Impacts of increased fertiliser prices in Malawi	227
6.6	Higher temperatures in the Sahel	230
6.7	The case of coffee	235
6.8	The downside of export horticulture	236
6.9	Production problems in achieving export quality	238
6.10	Foreign direct investment in the manufacture and retail of processed foods	242
6.11	Rethinking agricultural knowledge, science and technology	244
7.1	A view of sustainability as functional integrity and relative equilibrium	252
7.2	Sustainability as dynamic, complex and contested	253
7.3	Agricultural sustainability	257
7.4	Integrated pest management in rice	259
7.5	Fairtrade and market sovereignty	267
7.6	A typology of short food supply chains	275
7.7	School food in East Ayrshire	285
7.8	Community food security in action	287
7.9	Growing local food	289

Acknowledgements

I would first like to place on record my thanks to Janet Townsend for all her encouragement and support while I was in Cochabamba, Bolivia developing my understanding of, and appreciation for, the complex and dynamic interactions between food, agriculture, livelihoods and environment that inform much of this book. Subsequently, Wye College provided a valuable experience in working on a range of development and environmental issues, and I owe a great deal to Michael Redclift for sharpening my understanding of sustainability. I began seriously thinking about this book while a Visiting Professor at Colby College and I would like to thank Tom Tietenberg for his support while there. Some of the material set out here was first tested on students of the Master's in Food Culture at the University of Gastronomic Sciences in Italy and I would like to acknowledge their positive feedback. In Cork all my colleagues in the Department of Geography remain a remarkably good-humoured bunch of people considering the circumstances under which we labour, particularly since 2008. Denis Linehan deserves special mention for his advice and encouragement.

A lot of this book was written while I held a Senior Research Fellowship during 2008–09 awarded by the Irish Research Council for Humanities and Social Sciences. I would like to thank IRCHSS for their support, as well as the HEA for funding under PRTLI 3 which helped initiate some early thinking. I would like to thank Prof Anne Buttimer and, especially, Prof Jules Pretty for all their help and support. It has been a pleasure to collaborate with Mike Goodman over the last couple of years and Iain Farmer has been a great friend and long-distance walking companion.

Given the lapse of time, several editors at Routledge had the misfortune to keep a watching brief on my progress. While all deserve credit, Michael Jones did a great deal of prompting and prodding to good effect before he took to the hills and Faye Leerink has done a great job to get me over the finishing line. Throughout all of this Andrew Mould has been a stoic source of support. I would like to thank Michael Murphy in the Geography Department, UCC, for his excellent work on the figures. I would also like to thank the three anonymous referees who really did a thorough job in reading the manuscript and who lifted my spirits with their constructive and generous comments.

Permission to reproduce the following figures is also gratefully acknowledged: Cork University Press for permission to reproduce Figure 2.2; Earthscan Ltd for permission to reproduce Figure 2.3; and Elsevier Ltd for permission to reproduce Figure 5.3. I have not knowingly infringed any copyright on material reproduced in this book, but if there are any copyright holders who believe their work has not been adequately acknowledged they are asked to contact the publisher in order that any omissions or errors are rectified in future editions of this book.

Finally, I would like to thank Elmarie for her patience. This book is dedicated to my two brilliant and beautiful daughters, Liadán and Aisling.

CLS
Kilbrittain, West Cork

① Introduction

Why environment and food?

Eating, more than any other single experience, brings us into a full relationship with the natural world. The act itself calls forth the full embodiment of our senses – taste, smell, touch, hearing, and sight. We know nature largely by the various ways we consume it. Eating establishes the most primordial of all human bonds with the environment . . . (it) is the bridge that connects culture with nature . . .

(Rifkin 1992: 234)

No area of concern demonstrates the difficulty of managing the contradictions of the food system as clearly as the environment.

(Goodman and Redclift 1991: 202)

If, according to Wendell Berry, "eating is an agricultural act", then by extension it is also an ecological act. What we eat and how we eat has more impact on the Earth than almost anything else. Yet we have shown little interest in this connection until recently, when terms such as food miles, carbon footprint and Fairtrade have come into wider public discourse.

The basic biological necessity for all human life – the essential physiological requirement that we all share – is the need to eat. Historically, the kinds of food that we ate and the style in which it was consumed reflected a host of social, cultural and geographical factors. The past 100 years, however, have probably witnessed a greater transformation in the foods we eat than at any time since the Neolithic Revolution 12,000 years ago, when hunting and gathering gradually gave way to farming. During the twentieth century, significant changes gathered pace, initially in North America and Europe but spreading quickly to other societies around the world, involving a greater convergence in dietary practices with marked consequences for food production methods and for human health.

With respect to the former, the application of scientific methods and industrial technologies dramatically altered the scale and productivity of farming and food processing. Established practices in rearing animals and cultivating crops; the sourcing, processing and distribution of agricultural produce; and the purchase, preparation and consumption of food were all fundamentally transformed. This was to have a significant impact on the environment: on landscapes, soil and water resources, biological diversity and the global climate system. As regards human health, food is now cheaper in real terms and more readily available (for those with the means to purchase it) than ever before. But the ubiquity of highly processed, cheap and convenient foods is driving rising levels of diet-related diseases and associated risk factors (cardio-vascular, diabetes, obesity) on a global scale (Hawkes 2008).

We have arrived at a point where food has become a highly contested arena of competing paradigms (Lang and Heasman 2004). In the realm of production, some commentators insist that the "application of agricultural and food system science has been one of the great success stories of mankind". Fresco goes on to remind us that, since the 1960s,

> (W)orld population has doubled while the available calories per head increased by 25 percent. Worldwide, households now spend less income on their daily food that ever before, in the order of 10–15 percent in the OECD countries, as compared to over 40 percent in the middle of the last century. Even if many developing countries still spend much higher but declining percentages, the diversity, quality and safety of food have improved nearly universally and stand at a historic high.
>
> (Fresco 2009: 379)

These have been remarkable achievements to date, of that there is no doubt, particularly for those of us in the high-income countries of the world, as well as for increasing numbers of urban residents in middle-income countries who, on the whole, enjoy a privileged position within the global **agri-food system**. Clearly, science has played a major role in helping to feed the world, as the green revolution demonstrated (Godfray *et al.* 2010). However, if agricultural and food system science is such a success story, as Fresco argues, "because of the collective capacity of humankind to adjust to the lessons learnt", then this would appear to be a vital moment to take stock of the shortcomings and weaknesses of the current global agri-food system and remind ourselves of the following.

● Reliance upon the market as the sole effective mechanism for the supply and demand of food requires every individual to possess the capacity to buy. Such a mechanism cannot ensure equitable access to food, nor can it guarantee the provision of even basic nutrition for all. An estimated one billion people in the world are experiencing hunger and malnutrition because of their lack of **entitlements** through which to express a demand for food.

● In a context where international food policy appears no longer fit for purpose, the profit-seeking behaviour of food companies encourages them to promote those snack, convenience and confectionary products that are high in salt, sugar and fat. It has been suggested that over one billion people in the world are overweight or obese and susceptible to a range of diet-related diseases (Lang *et al.* 2009).

● The declining share of food in the household budgets of consumers does not reflect the true economic, social or environmental cost of its production, distribution and consumption. What we pay for food at the supermarket checkout does not take into account the loss of ecological services, the depletion of resources, the impairment of Earth system processes, and the rising medical costs of poorer human health.

● Calls for a doubling of food production to meet rising demand from a growing, more urbanised and possibly more affluent global population must not result in more of the same kinds of productivity-driven science and technology "solutions". Rather, questions of delivering global **food security** with sustainability will require new approaches that can ensure appropriate developmental, environmental and social justice outcomes (Pretty *et al.* 2010).

So, in a context where the world produces enough food for all at historically low prices, the global food system has created unprecedented numbers of underfed and overfed people. And at a time when it has recently been shaken by growing volatility of food prices, with severe impacts for the world's poor, it is confronted by a number of serious challenges, including the prospect of growing freshwater scarcity, a tightening of energy markets and greater climatic perturbance.

Food and the environment

There is now sufficient evidence and scientific consensus around the phenomenon of human-induced climate change, with warming of the climate system now regarded as unequivocal (IPCC 2007). According to a report produced for the European Commission, what we eat has more impact on climate change than any other aspect of daily life, accounting for 31 per cent of the **global warming potential** of products consumed

(Tukker *et al.* 2006). One category of foods that records the greatest environmental impacts across a range of categories is meat and meat products, with livestock production accounting for 18 per cent of global greenhouse gas emissions. Yet projections that talk of doubling demand for food by 2050 assume that consumption of meat will continue to rise sharply in rapidly developing economies as an inevitable aspect of the nutrition transition (Popkin 2006). With around 40 per cent of global meat production supplied by factory farms, increasing volumes of cereals and oil seed crops are being grown for livestock feeds, and account for at least one-third of total global grain output.

Agriculture accounts for around 86 per cent of global freshwater consumption and it has been estimated that one person alone might eat up to 5000 litres of "virtual water" per day, depending upon their diet, represented by the evaporation and transpiration associated with growing crops (Hoekstra and Chapagain 2007). Yet freshwater resources in some regions are seriously depleted, and raise questions about the appropriateness of water-intensive farming systems in dry countries producing export crops for distant markets.

Recent events have also revealed the degree to which the agri-food system has become entangled with global energy markets. In every aspect – from farming, through processing and manufacturing operations, to retail and to the point of consumption – the modern food system rests upon an abundant, low-cost supply of energy that is overwhelmingly provided by fossil fuels. Natural gas provides the feedstock for fertilisers, while oil is used to drive farm and other plant and machinery; is transformed by the petrochemicals industry into a wide range of packaging materials; and, critically, underpins the ever-lengthening supply chain that brings foods from all over the world to our local supermarkets. Concerns about rising energy prices and a belated realisation of the extent of our dependence on petroleum have resulted in a significant expansion of biofuel crops that compete with food crops for arable land.

Sustainability and food

These brief observations should indicate that the perspective taken in this book is a good deal less celebratory of agricultural and food system science than conveyed by the above quote from Louise Fresco (2009). This is not to deny the evident achievements of scientists *and* farmers to increase food output, but to make clear that we need a more holistic framework through which to evaluate the performance of the agri-food system than the adoption of singular "**productivist**" criteria such as output volume or yield. It is in this regard that the notion of sustainability has emerged as a framework capable of conveying important underlying principles across the biological, economic and social realms. One of the essential requirements of sustainability is to maximise goal achievement across these three realms through an adaptive process of trade-offs, a process that must necessarily be place- and time-specific. Consequently, it is not possible to reconcile sustainability with the kinds of scientific and technological "blueprint" created in the private or publicly funded research facilities of the North with the intention of rolling them out across the developing world. Striving to improve productivity is important, but so is achieving other system properties, such as stability (of output or prices); resilience and durability (abilities to withstand and recover quickly from such events as drought or crop pests); social justice and equity; and adaptation, among other things. At this stage, a detailed understanding of the notion of sustainability is not required – we will return to it at the beginning of chapter seven – but we can at least draw up some simple rules.

First, a **sustainable food system** must be able to demonstrate that it can optimise agricultural output without compromising the stock of natural resources and ecosystem services. However, it needs to move beyond that, extending through all subsequent stages of processing, manufacturing, distribution and retailing, to the point of consumption. Second, each stage of a sustainable food system must endeavour to minimise the use of non-renewable resources (such as fossil fuels), ensuring that the utilisation of renewable resources (such as wild capture fisheries) is within their capacity to regenerate stocks. Third, a sustainable system must reduce waste streams to a minimum and aim to bring pollution levels to within the capacity of ecosystems (the atmosphere, streams, rivers and estuaries) to deal with and neutralise these wastes (a process known as remediation). Finally, a sustainable food system is also committed to the principle of social justice, which means working to ensure the achievement of food – and, indeed, nutritional – security for all.

The notion of sustainability possesses markedly different meanings for different actors depending upon their position within the agri-food system. It is unlikely that a dairy farmer in Western Europe shares the same understanding as the CEO of a large food-manufacturing company, or that of a city-based office worker who buys their food from a supermarket, let alone an agricultural migrant worker toiling in the fields for below legal minimum wages: all will have quite different perspectives on what a sustainable food system looks like. Yet, if sustainability is to have any real meaning at all, that is, beyond the rhetoric of corporate "greenwash", then finding ways for farmers, farm workers and other food producers, as well as consumers and civil society actors, to have a voice in shaping the rules and principles of the food system will be vital.

It is apparent that, since the rise of the modern agri-food system after 1945 (discussed in chapter two), there has been a relentless squeezing of the public into the category of "consumers". Under the cover of a continuous and heavy barrage of advertising, consumers have had their attention refocused onto individualised concerns for convenience and low price. This has not only led to the exclusion from the mainstream of other food quality criteria (its sensual attributes, nutritional value, production methods, sourcing and traceability), but has allowed corporate interests to dominate the food system. This has left little room – at least until recently – for the majority of people to make more profound judgements about their food beyond narrowly circumscribed choices between competing brands. It has also created the legacy of a "knowledge deficit" in which scientists and policy makers assert a monopoly of wisdom over such things as genetically modified organisms on which the "average citizen" is regarded as uninformed.

With retailers especially anxious to resolve the contradictions of the food system on behalf of their customers ("How can meat be sold so cheaply if not produced under excessive stocking densities?" "Trust us: it's farm fresh quality assured"), we have come to take for granted so much of what our food system does. As beneficiaries of the in-store cornucopia, we have learnt not to ask questions, not to know too much, about the way in which our fillets of meat or green beans end up in polystyrene trays covered in a thin film of polymer. We are actively discouraged from peeking behind the curtain that conceals the production methods of the modern food system, and our sensibilities are easily disturbed by unwelcome news of practices associated with intensive animal farming, children in distant countries picking coffee or cocoa beans, or gangs of migrant labourers closer to home collecting field vegetables, salad crops or shellfish. While the contemporary food system frees most of us from toiling for our food beyond the task of pushing a trolley around the supermarket, it has also removed a significant degree of personal responsibility. We have learnt to accept the retailers' refrain that what we really want is cheap and convenient food.

Or have we?

There is increasing evidence that many people are choosing to take back some of their power from the corporate food system, to recover a degree of *food citizenship*. Developments, particularly throughout North America and Europe during the past two decades, have seen a variety of initiatives aiming to achieve a more sustainable food system characterised by a greater degree of **relocalisation**. There are many different aspects of this phenomenon, some of which are described in chapter seven, but among the most successful has been the growth of farmers' markets as sites for the retailing of ostensibly local products. Although there may be a number of different, locally contingent factors that underlie specific initiatives, there appears to be some common denominator around a desire for food that embodies different quality attributes (fresher, healthier, tastier); that is associated with a producer or someone who can speak on behalf of its authenticity; and that promises to carry less environmental impact. Above all, there has been a desire to recover food that is not only traceable and trustworthy, but also good to eat; food that is culturally, as well as physiologically, nourishing.

Focus and structure of the book

Before describing the organisation and structure of the book, I should explain something of its geographical focus. The intention in writing this book was to be as global as possible in its approach, reflecting the fact that the world today is criss-crossed in supply chains carrying **agri-commodities** from distant sites of production in poor rural regions to the supermarket shelves of affluent cities. In telling this story, there is potential for a great deal of "thick description", a narrative that would illustrate much of the detail of labour practices, production methods and consumption choices, but this is better done elsewhere (e.g. Cook 2004). Rather, the intention here is to outline some of the wider implications of this global system, and so specific examples here are, on the whole, briefer and more concerned with environmental and, potentially, systemic consequences.

Given that fish provide less than 1 per cent of the overall caloric intake of the world's human population, and less than 5 per cent of total food protein (Pimentel *et al.* 2008), it was decided to focus in this book on terrestrial – that is land-based – primary food production. Although aquacultural (fish farming) systems are growing rapidly in number and capacity, until recently fish consumption was largely met by harvesting wild populations (capture fisheries). The serious decline in wild fish stocks is now generally well known and represents a classic example of the unsustainable exploitation of a commons resource, where technological capacity (high-endurance factory ships) has outstripped biological capability (the reproduction of stocks). Although there is a brief illustration of aquacultural systems in chapter three, the scale and systemic importance of agriculture and its food derivatives remains the primary focus of this text.

Throughout the book, I generally use the terms North and South as shorthand to refer to the rich, powerful and generally liberal democratic states of what was once called the first world; and to those parts of the world once labelled the third world. I share the view of Williams *et al.* (2009), who note that the global South still remains something of a residual category that is often represented as the site of *problems*, especially poverty and environmental degradation. It comprises an increasingly huge diversity of countries at very different levels of economic development and prosperity, with a range of governance structures and political cultures, and an even greater variety of social organisation. In relation to the global agri-food system, however, we seek to understand the way in which countries of the South are connected to those of the North: through provisioning a widening range of food and feed commodities at low cost, and as sites where consumption is changing as a result of globalisation and the expansion of Northern-style dietary practices. The key throughout is a concern with relationality, the interconnectedness, of North and South.

While the ambition may be global, in reality a book is written from a particular geographical perspective, a result of one's own location and frame of reference. In this regard, and given its dominance here, the focus is inevitably concentrated on the global North, largely Western European, though with examples and evidence from North America, principally the United States, where possible. There is greater reference to examples from the United Kingdom than anywhere else, and this is deliberate for several reasons. First, there is the quality of documentation and data that exists for the UK. Second, the UK is, in global terms, a mid-sized society and economy of 62 million people with the capacity to exert a more influential global reach through the food supply chain than my own country of residence (with 4.2 million people). While the USA (population 250 million inhabitants) is more powerful still in shaping the global food system, it is more difficult to write authoritatively (and critically) about consumption practices of which one has less first-hand experience. Finally, as the first industrialised nation, the British diet remains a source of interest as it is said to have gone "straight from medieval barbarity to industrial decadence" (Goody 1997: 338). Industrial processes were applied to food earlier and on a larger scale here than anywhere else in order to achieve a cheap, mass-produced diet for the urban working class, the ingredients for which were sourced from all over the world. The earliest agri-food commodity chains mostly led to England.

And so to the structure of the book. Chapter two provides an overview of the global agri-food system. It begins by briefly tracing its recent development as a combination of global public policy and corporate interest. This has led to the formation of a liberalised market-driven structure in which countries are differentially integrated into a global food economy. Although countries of the global South have historically performed the role of producing a variety of tropical and subtropical agri-food commodities for consumers in the North, the chapter argues that their domestic markets have more recently become a focus for sales of processed food products. The global agri-food system is consequently a complex and dynamic system, exerting a different combination of effects in different places around the world. In order to understand better the relative importance of different aspects of the system, however, Figure 2.1 provides a schematic representation identifying four key elements: primary food production; the agri-technology industries; food trading, processing and manufacturing; and food retail and the food service industry. These elements are discussed in turn through the remainder of the chapter.

Chapter three is concerned with the resources and ecosystem services that support primary food production, principally agriculture. Drawing on the findings of the Millennium Ecosystem Assessment, it highlights the impact of our demand for food on the world's natural ecosystems, which perform vital regulating, provisioning and other services. Following a brief characterisation of the world's principal agricultural systems, the chapter proceeds to evaluate the state of four key resources required for food production: soils, water, biodiversity and energy. In relation to the first three, it is argued that their management under intensive farming practices has been marked by a degree of utilitarian complacency that has left these resources in a vulnerable state, albeit one that is still retrievable. As regards energy use, the rapid evolution of modern farming methods has been entirely underpinned by the availability of cheap fossil fuels. Signs pointing to a future of much higher energy prices suggest that urgent efforts will be needed to shift food production toward a more energy-efficient and certainly less fossil fuel-dependent path.

Chapter four maintains a focus on resources, and examines four key challenges, principally for primary food production. The first of these is climate change, which, as noted above in relation to the EU, is significantly affected by what we eat. Although the entire agri-food system is a source of greenhouse gases, the single most important element is agriculture, though this varies significantly between countries depending upon their mix of different farming (especially livestock) systems. The consequences of climate change will also have an impact on primary production in complex and differentiated ways, offering the prospect of short-term opportunities for some regions, but also the likelihood of medium- to long-term threats to global food security. For this reason, farming systems must at the very least find ways to reduce their emissions of greenhouse gases. The second challenge confronting primary food production concerns the stock of global freshwater resources. The extraction of water for irrigation is discussed through boxed examples illustrating how serious the consequences can be when resources are used at unsustainable rates. Through the application of the concepts of "virtual water" and "water footprints", the need to rethink more fundamentally patterns of production in contexts of relative water scarcity is explained. The third challenge is that of "peak oil", which argues that as we are at the midpoint of depletion in the total endowment of conventional oil and natural gas, it is likely that we face a future of declining supply. Indeed, it is only the development of non-conventional sources of fossil fuels and the expansion of biofuels that has served to dampen even higher increases in energy prices. Such a predicament represents a major challenge for primary food production. The fourth, and final, challenge discussed in chapter four concerns the rising demand worldwide for meat and livestock products. Why is this a problem? Principally because of the feed demands of intensive livestock production systems that account for around one-third of global cereal production, and the waste streams that result.

Chapter five moves beyond the farm gate and explores the elements downstream of agriculture: the processing and transformation of primary foods; and the manufacture, distribution and consumption of final food products and their environmental consequences. It makes clear that the development of those food products that make up the modern urban diet are the outcome of complex and interacting forces stretching back to the nineteenth century. The food industry has successfully built upon social and economic changes, as well as advances in technology, in order to produce an enormous range of products, to which thousands of new lines are added every year. The environmental dimensions of the many varied food processing and manufacturing operations around the world are outlined schematically, and again highlight the importance of water and energy to production. A section dealing with packaging reveals the wide range of materials needed to ensure the effective retailing of food and drink products, and once more the importance of oil, via industrial petrochemical processes, is made clear. We are reminded again of the powerful role that corporate retailers exercise, both in the way they have reconfigured supply-chain management in the sourcing of produce, and in its distribution to individual stores. Finally, the scandalous level of waste within the contemporary agri-food system is addressed. This is shown to comprise high levels of discard of perfectly acceptable product as well as the disposal of food regarded as surplus to requirements. With perhaps one-quarter of all purchased food thrown away, and with most affluent societies guilty of overconsumption, this represents a huge amount of resources utilised simply to exacerbate the waste management problem and increase the risk of greater ill health.

Chapter six focuses on the issue of global food security. It begins by tracing the evolution of the term, and raises questions about how effective an aspiration it is at a macro-level. The often-vexed issue of population growth is then addressed, noting how the clumsy Malthusian equation of numbers of mouths to volumes of food can be utterly misleading at a global scale. Given considerable regional differentiation as regards economic performance, together with a host of demographic variables, a much more spatially disaggregated analysis of food prospects is required. Such analysis would also need to take account of the changing global energy picture. Although the role of oil in food production has been discussed previously, its destabilisation of food security – as occurred during 2007–08 – is evaluated. The same must also be said for climate change: how this will affect different regions and countries around the world may also significantly affect their food security. Ultimately, however, this takes us back full circle to the issue of globalisation and the degree to which both production and consumption in distant countries are shaped by wider economic forces. The chapter closes with an evaluation of **food sovereignty**, a term that has come to recent prominence through the struggle of farmers and food activists in many countries around the world to assert their right to define their own form of farming and food consumption.

The last substantive chapter of the book, chapter seven, considers what a sustainable agri-food system might look like. Following a discussion of the notion of sustainability, in which the emphasis is placed on the need to appreciate the importance of complexity, dynamism and adaptation in socio-ecological systems, attention is turned to its application in the field of agriculture. Recognising that sustainable production methods would need to combine the particular social, ecological and technological resources and capabilities within an area, it is inevitable that this will result in a greater heterogeneity of agri-food systems. If production is therefore to become more diversified, this will have enormous repercussions for consumption; not least, it will encourage reflection on the moral basis of existing behaviours and ask how these might become more sustainable. Two issues are used here to illustrate the possibilities for more **sustainable consumption**: Fairtrade and meat eating. Following this, the chapter examines the potential for the relocalisation of food supply chains, and then concludes with a review of the possibilities for food planning. Combining mid-level government agencies with a revitalised notion of local citizenship, examples are used to demonstrate the capacities and potential for creating a more sustainable agri-food system.

Further reading

Counihan, C., van Esterik, P. (eds) (2008) *Food and Culture: A Reader*, 2nd edn. Abingdon, UK: Routledge.
Conscious that this present text entirely ignores the cultural dimensions of food in order to focus on its relationship with environment, this title is offered as an antidote. The thirty-six chapters here demonstrate and justify the cultural turn in food studies.

Lawrence, F. (2008) *Eat Your Heart Out: Why the Food Business is Bad for the Planet and your Health*. London: Penguin.
Guardian investigative journalist Felicity Lawrence regularly delivers some hard-hitting exposés of practices within the agri-food system. This collection highlights some of the most troubling.

Millstone, E., Lang, T. (2008) *The Atlas of Food: Who Eats What, Where and Why,* 2nd edn. London: Earthscan.
A valuable reference regarding the world of food through an excellent set of cartographic representations.

Patel, R. (2007) *Stuffed and Starved: Markets, Power and the Hidden Battle for the World Food System.* London: Portobello.
Roberts, P. (2008) *The End of Food: The Coming Crisis in the World Food Industry.* London: Bloomsbury.
These two titles are also good examples of the range of more popular and accessible texts that, like Lawrence's book, reveal some of the practices and resulting injustices of the global food system. Both bring together a huge array of stories and insights from the frontline with some insightful analysis.

② The global agri-food system

Contents

1. **Introduction**
2. **The development of the global agri-food system**
   Antecedents
   Theorising the globalisation of the agri-food system
   Corporate power
   Types of global integration
3. **The scale and structure of the global agri-food system**
   North–South dimensions
   Representing the agri-food system
4. **Primary food production**
   Technological change
   The productivist imperative
   The agricultural treadmill
5. **Agri-technologies**
   Mechanisation
   Chemical fertilisers
   Pesticides
   Seeds
6. **Food trading, processing and manufacturing**
   Exercising corporate reach
   The era of innovation
7. **Food retail and the food service industry**
   Food retailing
   Food service
8. **Summary**
Further reading

Boxes

2.1 Sweetness and sugar
2.2 King Sugar and other agricultural monarchs
2.3 Observations on the global dynamics of the agri-food system
2.4 The European agricultural revolution
2.5 Traditional low-input farming, pre-war UK
2.6 Features of the agricultural treadmill
2.7 The Green Revolution imperative
2.8 Stages in the process of motomechanisation
2.9 The concentration ratios for selected US food sectors
2.10 Some key features of the leading retailers

Figures

2.1 The global agri-food system
2.2 The farm as hub of the agri-food system
2.3 Stages in the process of motomechanisation
2.4 The egg-timer model of the food system

Tables

2.1 Global food sales 2002
2.2 IGD grocery turnover league

1. Introduction

There can no longer be any doubt today, at the turn of the twenty-first century, that we are in the midst of an unusually rapid change in all aspects of the world's agriculture–food system. This system consists of the farmers who produce the food, but also the huge industry that supplies farmers with inputs, from seeds to fertilisers to tractors to fuel, and the even larger industry that processes, packages, and distributes the food. And, although international trade in agricultural products has occurred for centuries, the pace at which the world is being bound together by trade and the penetration of third world agriculture by the largest of corporations is also quickening.

(Magdoff *et al.* 2000: 9)

Writing over a decade ago, Magdoff and colleagues highlighted not only the rapid pace of change, but the scale of production, degree of integration and extraordinary level of market concentration that was becoming ever more evident in the global agri-food system. During the past ten or so years, this process has intensified further, with the largest economic actors in the food system, most especially corporate retailers, extending their reach, power and influence. Today, the supermarket format has become a ubiquitous feature of food shopping around the world, from Auckland to Abuja, Beijing to Buenos Aires. Indeed, it is the cornucopia of the average suburban supermarket that best illustrates the extraordinary integration of the global agri-food system with extended supply chains, ensuring the year-round provision of fresh fruit and vegetables irrespective of their seasonality or the distance between their sites of production and consumption. Consequently, it is in large-scale retailing where most power in the food system now rests, reflecting the steady migration of "added-value" activities downstream and away from the farm-based sector during the course of the past century.

This chapter outlines the key components and dynamics of the agri-food system. It begins by discussing briefly how the process of globalisation was set under way, and how it has differentially transformed food and agricultural production possibilities around the world. The centrepiece of the chapter is Figure 2.1, which provides a schematic representation of the agri-food system. Four key elements are identified: primary production; the agri-technologies supply industry; food trading, processing and manufacturing; and food retailing and food service. Each of these is the focus of subsequent sections of the chapter.

First, however, we need to review briefly the development, scale and structure of the global agri-food system.

2. The development of the global agri-food system

Antecedents

For most of human history, long-range trade in food was limited to luxury items, with every society growing its own staples, at least until they could be imported cheaply. Where opportunities arose, an important motive for imperial expansion was the prospect of diversifying diets by imposing ecological collaboration on regions specialising in different foodstuffs (Fernandez-Armesto 2001). Elsewhere, trade was limited to high-value exotics that would enhance the flavour of staple foods, and therefore limited largely to the elite, at least until the seventeenth century. The exception to this was salt: a flavouring, but also essential to human life as well as being indispensable for food preservation. If salt could not be obtained locally then it quickly became the first and most important element of bulk trade.

> Where people ate a diet consisting largely of grains and vegetables, supplemented by meat of slaughtered domestic farm animals, procuring salt became a necessity of life, giving it great symbolic importance and economic value. Salt became one of the first international commodities of trade; its production was one of the first industries and, inevitably, the first state monopoly.
>
> (Kurlansky 2003: 11–12)

The international trade in spices can be traced back thousands of years, with Indian pepper the most commercially important. Travelling overland it moved westward along the Silk Roads of Central Asia, or by sea it was shipped across the Indian Ocean to ports along the Persian Gulf, from where Arab merchants handled trade as far as the Mediterranean. Given the distance and number of hands through which the spices passed, it was little wonder that Europeans paid so much for them. Although there were disruptions to trade, as civilisations *en route* flourished and faded, by the early medieval period trade from the east eventually helped Italian city-states such as Genoa, Florence and Venice become major commercial centres. With Arab traders now voyaging to the islands of Sumatra, Java and the Moluccas, cloves, nutmeg and mace now found their way to the West. Yet by the fifteenth century, with Constantinople falling to the Turks and pepper prices in Venice up by a factor of thirty, it became clear that Europeans needed to find their own way to the Indies (Davidson 2006). The "Golden Age" of European navigation and "discovery" – regarded in the South as the onset of European conquest, imperialism and plunder – had begun.

According to Fernandez-Armesto (2001), there were three stages in the great change that converted the traditional world of the eastern spice monopolies into a global system. First, the westward transfer of the world's main centres of sugar production, which started in the late Middle Ages; second, the development in the sixteenth and seventeenth centuries of new trade routes, over which western merchants enjoyed privileged access; and, third, the progressive control over production by western powers deploying violent methods. Although all three stages are worthy of attention, it is sugar that best illustrates the way in which production of one food commodity could have such enormous ramifications for a large part of the world. Although not a spice, sugar was an exotic condiment that could be obtained only by trade at great cost; moreover, unlike pepper, cloves and cinnamon, sugar offered Europeans the chance to grow it themselves. In order to understand the driving force behind this geographical shift in production, Box 2.1 summarises Mintz's (2008) account of the rising consumption of sugar from the early medieval period.

Box 2.1

Sweetness and sugar

Until the seventeenth century, ordinary folk in Western Europe secured sweetness in food mostly from honey and fruit. Though sugar cane was grown in South Asia as early as the fourth century BC, evidence of processing – boiling, clarification and crystallisation – dates from almost a millennium later. And although sugar cane was being produced on the southern littoral of the Mediterranean by the eighth century AD, during those centuries it remained costly, prized, and less a food than a medicine. It appears to have been regarded much as were spices, and those who dealt in imported spices dealt in sugar as well. By the thirteenth century, English monarchs had grown fond of sugar, most of it probably from the Eastern Mediterranean. In 1226, Henry III appealed to the Mayor of Winchester to obtain for him three pounds of Alexandrine sugar; by 1243, when ordering the purchase of spices for the royal household, Henry III included 300 pounds of *zucre de Roche* (white sugar). By the end of that century, the court was consuming several tonnes of sugar a year, and early in the fourteenth century a full cargo of sugar reached Britain from Venice. By the fifteenth century, sugar was beginning to reach England from the Atlantic plantation islands of Spain and Portugal, and by now had begun to enter into the tastes and recipe books of the rich.

Although there is no generally reliable source upon which we can base confident estimates of sugar consumption in Great Britain before the eighteenth century, there is no doubt that it rose spectacularly. One authority estimates that English sugar consumption increased about fourfold in the last four decades of the seventeenth century, and trebled again during the first four decades of the eighteenth century; it then doubled again from 1741–45 to 1771–75. English and Welsh consumption increased about twenty times in the period 1663–1775. Since population increased only from 4.5 million to 7.5 million, the per capita increase in sugar consumption appears dramatic. By the end of the eighteenth century, average annual per capita consumption stood at 13 pounds; the nineteenth century showed equally impressive increases; and the twentieth century showed no remission until the past decade or so.

Sugar consumption in Great Britain rose together with the consumption of other tropical ingestibles, though at differing rates for different regions, groups and classes. This general spread (of tea and coffee as well as sugar) through the western world since the seventeenth century has been one of the truly important economic and cultural phenomena of the modern age. These were, it seems, the first edible luxuries to become proletarian commonplaces; surely the first luxuries to become regarded as necessities by vast masses of people who had not produced them; and probably the first substances to become the basis of advertising campaigns to increase consumption.

Source: adapted from Mintz (2008), 92–3.

Box 2.2

King Sugar and other agricultural monarchs

The demand for sugar produced the plantation, an enterprise motivated by the proprietor's desire for profit and placed at the service of the international market Europe was organising. The plantation was so structured as to make it, in effect, a sieve for the draining-off of natural wealth. Each region, once integrated into the world market, experiences a dynamic cycle; then decay sets in with the competition of substitute products, the exhaustion of the soil, or the development of other areas where conditions are better. The initial productive drive fades with the passing years into a culture of poverty [. . .] The Northeast was Brazil's richest area and is now its poorest; in Barbados, Haiti and the many other "sugar islands" of the Caribbean, human ant heaps live condemned to penury [. . .]. And this has not been the role of sugar alone: the story has been the same with cacao [. . .] with the spectacular rise and fall of cotton in Maranhão; with the Amazonian rubber plantations; with the fruit plantations in Brazil, Colombia, Ecuador, and the unhappy lands of Central America. Each product has come to embody the fate of countries, regions and peoples [. . .] The more a product is desired by the world market, the greater the misery it brings to the Latin American peoples whose sacrifice creates it.

Sugar had destroyed the Northeast. The humid coastal fringe, well watered by rains, had a soil of great fertility, rich in humus and mineral salts, and covered by forests. Naturally fitted to produce food, it became a place of hunger. At the end of the sixteenth century, Brazil had no fewer than 120 sugar mills [. . .] but their masters, owners of the best lands, grew no food. They imported it, just as they imported an array of luxury articles which came from overseas with the slaves and bags of salt. Abundance and prosperity went hand in hand, as usual, with chronic malnutrition and misery for most of the population.

Source: adapted from Galeano (1973), 72–5.

By the fourteenth century, sugar cane's westward march took in Cyprus and Sicily, from where it was taken to the Algarve in Portugal and then to the Atlantic islands: Madeira, the Canaries and the Cape Verde islands. It was very quickly transplanted to the New World following the Conquest: indeed, sugar might be regarded as one of the principal instruments of domination. By 1513, the first sugar mill opened on the island of Hispaniola (present-day Haiti and the Dominican Republic), where Columbus had arrived for the first time in 1492 (in search, it might be recalled, of a more direct route to the Indies and the source of those expensive spices). By the 1530s, cane plantations were being established in the north east of Brazil and then proceeded to utterly dominate human life in this region and beyond for centuries. Though underpinned by the apparently insatiable demand for its consumption, the advance of "King Sugar" was driven by an even greater avarice for the profits that it realised: certainly by Portuguese and Spanish plantation owners, but especially by Dutch and British interests, who supplied the labour and refined the crude sugar. For the key to its production was trans-Atlantic slavery: an estimated 5–6 million Africans were transported to Brazil alone up until the nineteenth century; perhaps 10 million crossed the Atlantic in chains over three centuries. The consequences of sugar for the New World – and for Africa – have been devastating and long lasting. Box 2.2 provides an extract from Eduardo Galeano's book about Latin America that helps us to understand better that such developments were not a series of unfortunate events, but part of systematic and organised plunder directed by distant elites that extend beyond the history of sugar.

Theorising the globalisation of the agri-food system

"You believe perhaps, gentlemen," said Karl Marx in 1848, "that the production of coffee and sugar is the natural destiny of the West Indies. Two centuries ago, nature, which does not trouble herself about commerce, had planted neither sugarcane nor coffee trees there."

(Galeano 1973: 77)

It is apparent that, although we might wish to focus on the contemporary dynamics of the global agri-food system and their consequences for the environment, it is vitally important to trace its historical development and critically evaluate the driving forces that have shaped it. Since Marx, numerous scholars of many disciplines have invested enormous effort in devising theoretical models and conceptual tools that would help us not only to understand the past better, but also to aid in explaining the unfolding circumstances of the present and future. While the scope of this book largely precludes us from visiting many of the debates that have centred on global food politics and agrarian structures, it is important to recognise how central food has been in establishing global divisions of labour and maintaining the enormous disparities of wealth, power and opportunity that result. So, as Marx observes, the planting of sugar cane and coffee in the Caribbean was not a "natural" event, but has served effectively to lock much of the region into the export of agricultural commodities, with the result that its social and economic progress has been slow, uneven and largely unsustainable. Consequently, while this book will have to skirt around important debates regarding the macro-economic structures that have shaped the global food system, this is in the interests of teasing out the environment–food linkages – not because these debates are irrelevant or uninteresting. On the contrary, as Friedmann observes: "Food and agriculture are superb lenses into tensions in the larger political economy because they express the contradictory commodification of the human (physiological) and natural (land) foundations of society" (Friedmann 2009: 337). Interested readers might follow up on suggested further reading listed at the end of this chapter.

One issue that has generated considerable debate concerns the more contemporary issue of globalisation. For example, it has been argued by some that the deepening and accelerating development of the global agri-food system is simply a logical progression of long-established international supply chains that can be traced back to almost the earliest phase of European expansion in the sixteenth century. In this view, one of the goals of conquest, imperial subjugation and, later, colonial control was to achieve a supply line of foodstuffs for consumption in the metropolitan powers. Others, however, reject this increasing global integration as a logical continuation of the past, arguing that the recent era marks a new and profound shift in the development of the global agri-food system. There is also widespread disagreement about whether the term "globalisation" is helpful in either describing or explaining this rapid internationalisation of food supply chains.

Yet globalisation has become a widely used term that conveys the multiplication and intensification of worldwide social, economic, political and cultural relations. In harness with the emergence of new communication and media technologies and practices, which have smoothed the friction of distance and time, globalisation has been represented as evidence of greater economic convergence toward a common destiny. Yet, while globalisation can on occasion be a useful shorthand term, it is also true to say that it is frequently presented as an inevitable process ("the end of history"), which will sooner or later either make us all beneficiaries of its ineluctable economic logic, or otherwise the helpless victims of its voracious greed. During the course of this book, it will become apparent that the economic transformations wrought by globalisation are more incomplete, discontinuous and uneven in their consequences and effects than is often presented. First, though, it is necessary to untangle the driving forces behind it.

Although responsibility for globalisation is commonly attributed to transnational corporations, it is important to understand the measures taken within the realm of global governance, by multilateral agencies in consort with powerful national political interests, that shaped the conditions for globalisation to unfold. The 1970s was a decade marked by leveraged increases in oil prices and the consequent lending of surplus petro-dollars by commercial banks. When this was followed by a steep rise in interest rates accompanied by global recession during the early 1980s, a number of states, including the largest in Latin America, were in danger of defaulting on their debt repayments. This led to the intervention of the International Monetary Fund and the World Bank, which established structural adjustment programmes designed to ensure that governments continued to meet their international financial obligations. One of the key measures of the structural adjustment programmes, alongside cutting public expenditure and privatising state assets, was to maximise export revenues. Consequently, a new era of export orientation was initiated.

At the same time, the rolling out of the liberalisation agenda, best exemplified by the conclusion of the Uruguay round of trade talks in 1994 that led to the creation of the World Trade Organization (WTO), heralded a process whereby states were expected to open up their markets to both imported goods and direct foreign investment. Although agriculture had long been a contested issue during the Uruguay round, and continues to be so in all subsequent WTO summits – not least because the United States and European Union both maintain highly protected and heavily subsidised farm sectors while seeking to open up developing country markets – the liberalisation agenda has prevailed. The first principle of this agenda is that of comparative advantage, derived from the nineteenth-century political economist David Ricardo, who argued that a country would be better off by specialising in the production of things it was best at, rather than producing everything it needs. This principle was expressed very clearly in 1986 by then US Agriculture Secretary John Block, who observed: "the idea that developing countries should feed themselves is an anachronism from a bygone era. They could better ensure their food security by relying on U.S. agricultural products, which are available in most cases at lower cost" (McMichael 2005: 278).

With the successful conclusion of the Uruguay round, requiring the removal by developing countries of tariffs and other protective measures to defend their farmers, large volumes of low-cost surplus grain from Northern producers were released onto Southern markets, driving down agricultural commodity prices as a consequence. In the EU, the Common Agricultural Policy's McSharry Reforms of 1992 replaced farm price supports with direct payments and sought to introduce world market prices, while maintaining compensatory measures to buffer European farmers. In the South, there was no such protection and, with the elimination of "trade-distorting" measures, increased flows of low-cost raw materials began.

Corporate power

This, however, is precisely the point where transnational private companies enter the picture. Large, long-established companies such as

Cargill, ConAgra and ADM, that were already dominant in the international buying, shipping and selling of grain, were well placed to take advantage of these new open markets. Besides those companies with established globally branded goods, many other agri-food companies had hitherto been principally focused on regional markets up to the 1980s. With the onset of trade liberalisation, however, there were very rapid developments across the entire agri-food system: agri-technology suppliers built local supply chains to deliver the large volumes of chemical fertilisers and pesticides required by countries to improve export production; food processors and product manufacturers established a presence for their globally branded soft drinks, snacks and other convenience food products (often through merger and acquisition of local companies, then encouraging shifting allegiance from existing national brands to "cooler" western-style products); fast food franchises, of course, have been at the vanguard of cultivating new consumer tastes; while supermarkets arrived later but are probably becoming the most influential sector within the international agri-food system.

Globalisation of the agri-food system is not simply about extending the reach of private sector actors, or measured by rising volumes of internationally traded goods, however, but embodies the creation of new norms and standards, regulations and institutions, consumption patterns and scientific practices. von Braun and Diaz-Bonilla (2008), for example, refer to the "science, knowledge, and information content of the agri-food system becoming increasingly internationalised" as evidence of globalisation. While this normative observation may be correct, the increasing application of science and technology developed for the purpose of maximising agricultural output rather than meeting local food and livelihood needs comes at a cost. Western scientific knowledge invariably displaces empirically derived and context-specific local knowledge that has evolved over generations and which has served to sustain livelihoods in often risk-prone environments. International knowledge comes laden with intellectual property rights, for example, patented seeds that serve to extend the control of transnational seed companies over the interests of farmers.

In this respect, the internationalisation of food supply chains has developed and deepened as capital has embarked on new rounds of investment, taking advantage of cheap labour and ecological resources in order to facilitate production opportunities destined to supply distant markets. Although part and parcel of the prevailing economic paradigm of neoliberal free trade, with its commitment to cost savings and price reductions in order to undercut competition and increase market share, this internationalisation is also represented as evidence of a "pro-poor" development agenda, where an emphasis on increasing exports of agricultural products from poor countries into global markets demonstrates a commitment to generate jobs and improve income while reducing poverty. Unfortunately, the difficulties of reconciling these two contrasting agendas (a "race to the bottom" in pursuit of the lowest prices on the one hand, a desire to generate wealth on the other) means that it is not always clear if the principal beneficiaries are the poor, in whose name such schemes are justified, or local elites and their overseas buyers, where value seems to accumulate. This is a question worth asking when reviewing three different ways in which countries of the South have experienced closer integration into the global agri-food system.

Types of global integration

The first type of integration into the global agri-food system really has to begin with that which best typifies the long-standing trading relationship for countries of the South and those of the North: the production and export of tropical agricultural commodities such as coffee, tea, cocoa, sugar cane, cotton and bananas. Given the strong encouragement by the international financial institutions during the 1980s, countries made substantial efforts to increase exports of these primary commodities, often in a minimally processed state. Yet, faced by inelastic demand (we can only drink a certain amount of coffee); competition from industrial substitutes (e.g. cane sugar has faced the challenge of synthetic sweeteners and high-fructose corn syrup, as well as sugar beet); and structural overproduction (continuous excess supply), prices fell sharply. According to Weiss (2007), the internationally traded price of tea had fallen to 47 per cent of its 1980 value by 2002, groundnuts to 38 per cent, palm oil to 24 per cent, cotton to 21 per cent, cocoa to 19 per cent and coffee to just 14 per cent. Yet there remain today around forty countries for which the proceeds from the export of just one of these commodities still account for more than 20 per cent of their total merchandise earnings. It demonstrates, however, that despite a "pro-poor" discourse, the international agri-food system's encouragement to produce more tropical agricultural commodities is not truly aimed at improving the economic prospects of the poorest, that is, unless there is a Fairtrade or other premium attached (to be discussed in chapter seven).

A second and more recent type of integration into the global food system has been the development of high-value horticulture and aquaculture. By the mid-1990s, around twenty or so lower-middle-income countries accounted for the export of US$500 million of high-value foods each year. One of these countries, Kenya, is considered a real success story, having witnessed considerable growth in horticultural exports amounting to $350 million by 2003, which was divided between fruit and vegetables on the one hand and cut flowers on the other. Kenya has a number of strategic assets that have enabled it to take advantage of the changing agri-food system, among them location; an existing tourist infrastructure with established air links to Europe (initially horticultural exports could be shipped on passenger flights, until such time that export volumes made cargo planes viable); Asian entrepreneurs with well developed networks; and a history of settler agriculture and expatriate landholding. By 2007, horticultural and cut flower exports provided well over US$1 billion to Kenya's economy, with flower exports alone earning US$600 million, an 80 per cent increase on 2006 (New Agriculturist 2009). Other countries that have benefited from the development of high-value exports include Brazil, Mexico, Chile and South Africa.

A third way by which some countries of the South have become increasingly integrated into the global agri-food system has been through the development of large-scale agro-exports capable of competing on international markets with the very largest and lowest-cost producers, such as the USA and EU. Here, Brazil has probably achieved pre-eminence, producing more coffee, sugar cane, cassava, bananas and sisal than any other country, and ranking second in the world in the production of oranges, cocoa and soybeans. It is also a major exporter of beef and poultry. In the early 1970s, Brazil's agricultural economy was poorly developed, largely dominated by sugar cane (since the mid-sixteenth century) and coffee. Yet, in response to its severe debt burden and the imposition of structural adjustment measures, Brazil underwent a profound process of agro-industrialisation. As is well known, an important aspect of this was the rapid expansion of agriculture into Amazonia, with the forest cleared to make way first for cane, then cattle, and more recently soybeans as the frontier has moved steadily forward. Besides the enormous ecological consequences, an issue to which we return later, this agro-industrial model has exacerbated significant social inequalities. For example, an estimated 5 million rural families lack access to land, and a quarter of all Brazilians live on less than $1 a day, while 3 per cent of the population control two-thirds of all arable land.

Consequently, to speak of a global agri-food system is not to suggest that it has become a vertically integrated, corporate-controlled, universal model characterised by the kinds of flexible specialisation that distinguish the computer or car manufacturing industries. If globalisation is at best partial and contingent, such that many of the poorest countries are largely excluded from the benefits of participation in a world economy, this is especially so with regard to the agri-food system, where ecological circumstances, lack of infrastructure and limited attractiveness for investment might render such regions invisible to corporate interests. It is conceivable that such regions might still possess some limited autonomy to sustain endogenous food production systems in line with local biological possibilities. Elsewhere, the reorganisation of local production systems into platforms for lucrative agro-exports demonstrates the uneven and discontinuous process of transformation that can leave areas potentially vulnerable to global financial fluctuations and local ecological perturbance. Box 2.3 provides some observations on the consequences of the globalised agri-food system.

Box 2.3

Observations on the global dynamics of the agri-food system

For Heasman and Mellentin, the modern global food system has:

- witnessed a massive increase in food supply regardless of broader human and environmental health aspects, the economic costs of which are "externalised";
- become dominated by certain grains (wheat, maize, rice), and livestock production which promotes meat and dairy product consumption;
- seen the intensification of agriculture and chemical use with a tendency towards larger production units and fewer crops and farmers;
- involved costly farm support measures, in the form of subsidies, in the trade-dominating blocs, often at the expense of smaller producers and rural communities and alternative uses of public monies;
- distorted markets and prompted unequal and unfair trade, mainly to the detriment of poorer countries;
- created a culture of food dependency in developing countries, characterised by "food aid" and food imports from rich producers, and the

setting up of domestic production in poor countries for the export markets of the rich food shopper;
- seen increasing national, regional and global restructuring by large food business and its associated supply industries, built around a select number of commodities;
- seen environmental concerns (such as falling water tables, reduced biodiversity, soil erosion, chemical contamination and disposal of animal wastes) become major problems.

Source: adapted from Heasman and Mellentin (2001), 28.

3. The scale and structure of the global agri-food system

Estimates put the aggregate value of the global sale of food in the early 2000s at US$4 trillion each year ($4 \times 10^{12}$) (Gehlhar and Regmi 2005). This was probably a significant underestimate, as such a figure does not take account of food produced and consumed by farm families, or products that make their way through more informal supply channels. So, while it calculates the value of food and drink sales in supermarkets, bars, restaurants and fast food enterprises around the world, it fails to acknowledge the importance of traditional markets or street vending activities. Moreover, it is apparent that large industrial-scale food processing and manufacture are growing at such a rate – together with supermarket retailing and the food service sector – in countries of the South, that by 2011 the true estimate may be closer to twice the above figure.

Table 2.1 provides a broad breakdown of global food sales between retail stores and food service sectors. The latter refer to all establishments selling food and drink that is consumed outside the home; while the retail category encompasses a range of different operations stretching from the largest Walmart hypermarket to the small independent grocery shop. Processed food products is a term used to describe materials that have undergone a transformation from their raw forms, either to extend shelf life (e.g. freezing or dehydration of fruit and vegetables) or to improve consumer palatability of raw commodities (e.g. transforming grains into bakery products, or livestock tissue into meat products). Packaged food products, meanwhile, are those sold through retailers as prepared foods for home use or for direct consumption.

Table 2.1 *Global food sales 2002*

Food type	Retail stores	Food service	Total
Fresh food	531	382	913
Processed products:	1762	1420	3182
packaged food	1148	828	1976
beverages	614	592	1206
alcoholic drinks	316	422	729
hot drinks	53	12	65
soft drinks	245	167	412
Total food	*2293*	*1803*	*4096*

Source: Gehlhar and Regmi (2005).

North–South dimensions

As the data in Table 2.1 indicate, processed products account for three-quarters of total world food sales, and of this an estimated 60 per cent takes place in the North. Gehlhar and Regmi (2005) explain how such products already effectively saturate these high-income markets, offering limited growth potential in the future. In developing country markets of the South, however, there has been very rapid growth in the sales of packaged foods, given the limited penetration of those markets to date, recent improvements in per capita income, and higher rates of population growth. Drawing on survey evidence from Euromonitor, a market analysis company, Gehlhar and Regmi review the average annual rate of growth in sales volume over the period 1998–2003 for selected food categories. Sales of breakfast cereals, for example, comfortably achieved double-digit rates of growth in low- and middle-income countries (India 11.7, Romania 28, Mexico 15 per cent), while the UK and US markets recorded growth of 1.0 and 1.3 per cent, respectively. Similar patterns were recorded for other categories of food: for example, oils and fats, the market for which contracted in Germany (by 1.7 per cent) and the USA (by 0.1 per cent), whereas it grew rapidly in Brazil (by 24 per cent) and Romania (by 29 per cent); and dairy products, where the market is largely flat in the North (UK 1.8 per cent), whereas it is expanding quickly in China (by 15 per cent), Colombia (by 13 per cent), and Indonesia (by 15 per cent) over the period 1998–2003 (Gehlhar and Regmi, 2005).

Unsurprisingly, there has been considerable effort by the major food manufacturers to "build their brands" in the South, which means using advertising and promotion to encourage people to switch from traditional foods to these new packaged products. Although this can involve increased trade in processed food products, critically much of the recent growth is largely due to the establishment of manufacturing operations by foreign transnational corporations in countries of the South. Commodity-based products are those most closely linked to a specific agricultural commodity, such as meat, fruit and vegetables, fish or milk; processing and packaging tend to take place close to the location of primary production and the products are then exported (as frozen meat, canned fruit or fish, or dried milk powder). Such products account for 75 per cent of US food exports. Yet manufactured food products that draw together a variety of ingredients for combining into consumer-ready branded packages are more likely to be manufactured in the country of consumption, as they can be tailored to local tastes and preferences.

Consequently, this sector has seen significant foreign direct investment (FDI) through either the creation of an independent company "branch plant" or the acquisition of, or merging with, an established local company. Gehlhar and Regmi (2005) note that processed food sales in 2002 by US companies through FDI (i.e. by creating local manufacturing operations) amounted to US$150 billion, or five times the value of product exports from the USA. By 2006, FDI by US food-manufacturing companies amounted to US$67 billion, although the beverages sector accounted for over half (US$35 billion) of this. Note that a further $16.4 billion of US FDI was directed toward the establishment of food retail and food service activities. In other words, building processed food and drink manufacturing facilities, and creating networks of supermarket retailers and fast food outlets overseas, now constitutes a bigger and more profitable strategy than the export of processed food products from the United States.

Representing the agri-food system

It should be becoming apparent that the global agri-food system is a complex and dynamic system that is having a major effect all around the world, though in some places with greater consequence than in others.

While trade has grown significantly between countries, much of the transformation in patterns of consumption and in specific sites of production is due to the activities of transnational companies. Figure 2.1 provides a schematic representation of the global agri-food system. It comprises five key elements: the agri-technologies supply industry; primary production; food trading, processing and manufacturing; food retailing and food service; and consumers. These elements, with the exception of the latter, are dealt with in turn throughout the rest of this chapter.

The elements of Figure 2.1 might suggest a very linear and ordered way in which the flow of food from primary producer passes through processor and retailer stages, ultimately arriving in the consumer's kitchen. It would be incorrect to assume, however, that each component of this system operates with relative autonomy – as will become clear, the exercise of power and control over food can extend beyond the boundaries of each element, and particularly in an upstream direction. For this reason, I have avoided using the term "stages", which conveys an overly simplified impression of a discrete and logical progression of food. The configuration of ownership and control over a particular supply chain will depend on the commodities and market conditions involved, as, for example, with the integration of the grains–feed–livestock complex.

Several of the boxes feature the names of leading businesses within that part of the food system, although this does not mean that their interests are confined to that element. The high degree of concentration in agri-technologies or food retail stages contrasts sharply with primary food production, where there may be up to 450 million farms, the vast majority of them smallholdings of less than 2 ha (von Braun 2005). Besides the five main elements, the entire agri-food system is clearly underpinned by a very wide range of ancillary industries, services and institutions. These include manufacturers and suppliers of machinery, equipment, materials, energy, advertising and public relations; financial services, providing a range of functions, from the provision of credit and insurance against loss, to fostering more speculative activities such as trading in commodity futures; and "food governance", encompassing the regulation of food standards, from the global level by the Codex Alimentarius through to national food safety authorities. Finally, the agri-food system is underpinned by continuous research and development conducted, for example, by publicly funded, university-based scientists, as well as those working in the laboratories of the major food companies themselves, although the distinction between

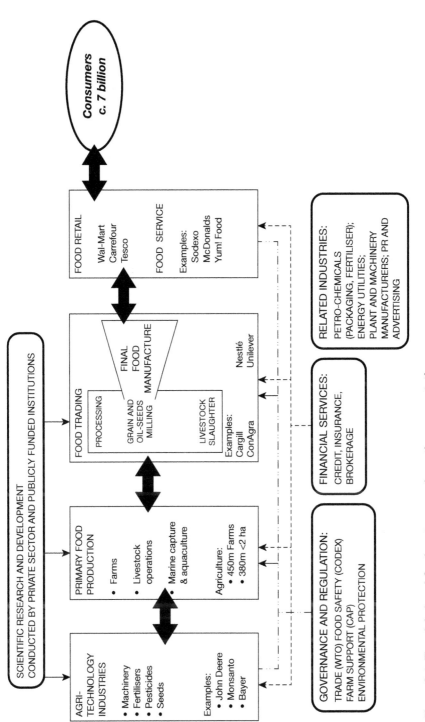

Figure 2.1 The global agri-food system: a schematic representation

them is becoming increasingly unclear as university facilities and personnel are used for product development.

Figure 2.1 is an effort to capture schematically this range of actors associated with the agri-food system, and provides a point of reference throughout the book. For the remainder of this chapter, each of the elements is considered in turn, in order to understand the role it plays within the overall agri-food system, how this role has changed in the modern period, and how it has been shaped by the other elements. We begin with the primary food production component.

4. Primary food production

The box labelled "primary food production" in Figure 2.1 refers to the production of materials that derive from the management of biological processes and that yield foods consumed in their natural state (fresh fruit and vegetables, for example) or that become ingredients for further processing and manufacturing into consumer products. As humans obtain over 99 per cent of their calories from the land, discussion here concentrates on terrestrial production, especially modernised agriculture across the North and South, which is most closely integrated into the global agri-food system. The purpose of this section is to understand how agriculture became increasingly squeezed "between a set of extractive, mechanical, and chemical industries located upstream", supplying it with inputs and means of production, and "a set of downstream industries and activities which stock, process, and sell its products" (Mazoyer and Roudart 2006: 376). While this process became especially intensified during the second half of the twentieth century, it is useful to remind ourselves that there was an earlier Agricultural Revolution across Europe, which had not only introduced a range of technological innovations, but consolidated the privatisation and enclosure of common land. Box 2.4 explains.

Technological change

Mazoyer and Roudart's historical review explains how, from the first half of the nineteenth century, a rapidly growing iron and steel industry began to produce all sorts of new equipment for agriculture and for transportation. Initially, such equipment was designed for animal traction but, for example, the traditional wooden-framed plough inherited from the Middle Ages was given over to all-metal construction. In the United

---

Box 2.4

The European agricultural revolution

During the late 17th, 18th and early 19th centuries, there was a period of extraordinary innovation in agriculture in Europe – so much so that this is now known as the Agricultural Revolution, as if it were the only one, rather than just the latest before our modern period. Over a period of about 150 years, crop and livestock production in the UK increased three to fourfold, as innovative technologies, such as the seed drill, novel crops such as turnips and legumes, fertilisation methods, rotation patterns, selective livestock breeding, drainage, and irrigation, were developed by farmers and spread to others through tours, open-days, farmer groups, and publications, and then adapted to local conditions by rigorous experimentation.

Source: Pretty (2002), 31.

States, John Deere began the industrial manufacture of iron and steel ploughs that sold by the hundreds of thousands. Other equipment soon followed, with genuine mechanical innovations that saved significantly on labour time: harrows and rollers for soil preparation; sowers, hoeing machines and ridgers for seeding, weeding and soil mounding; and reapers, binders and threshing machines for cutting grass and grain, gathering stalks into sheaves, and removing grain from the stalk. Other hand-cranked machines were developed, capable of winnowing the grain from the chaff, cutting roots for animal feed, churning milk for butter and cream, and so on. All these innovations constituted the final improvement of largely animal-drawn cultivation in the industrialised countries, reducing the agricultural labour force by half.

Such innovations were most quickly adopted in the United States, and then by farmers in other white settler countries – Canada, Australia, Argentina – which shared the characteristics of larger average farm size and a relative scarcity of labour. In Europe, adoption was slower, given that the average farm could make this new equipment cost effective only if they were to reduce the family labour force or expand the cultivated area, which inevitably, given the lack of available land, would be at the expense of other farms. Although this varied between regions, the mechanisation of animal traction was still occurring through the first half of the twentieth century, with artisan craftsmen (blacksmiths, cartwrights, millwrights) manufacturing and maintaining vernacular technologies for local farms. Note that the displacement of animal traction by motorised forms (tractors, harvesters) only began to get under way during the inter-war period, a process that was rapidly stepped up after 1945.

In the second half of the nineteenth century, the steam engine made a significant contribution to an agricultural revolution, not primarily in the performance of field tasks (although large steam engines maintained by enthusiasts are a feature of many rural fairs today, their applications at the time were largely confined to threshing), but in transportation. The development of steamships with screw propellers first allowed fertilisers and soil amendments to be transported over much greater distances, for example, nitrates from northern Chile and guano from Peru. Secondly, it allowed agricultural products themselves to be moved in bulk and quickly between continents. As American wheat production (US, Canadian and Argentinean) grew during the latter part of the nineteenth century, larger volumes were imported across the Atlantic, arriving in Europe at prices below the costs of production here. By the turn of the century, European agriculture was in crisis, though each country developed different strategies to address this. In the UK, where a free trade doctrine prevailed – naturally, given the export of its manufactures around the world – the farm sector was deeply affected; other countries practised some form of protectionism (e.g. France) or developed specialisation in perishable products (e.g. Denmark, the Netherlands).

At the turn of the twentieth century, then, across most of Europe, the agricultural landscape was marked by diversified crop and livestock husbandry. There was far from universal adoption of the new animal-drawn mechanical equipment, and large areas still relied on tools of medieval origin, such as the ard (a simple plough) in the Mediterranean region. As a consequence, farms tended to produce a large variety of products for direct consumption in addition to cash sales of surpluses with which to buy tools and other essential products, and pay taxes, rents and tithes. Depending on the region and the possibilities of production, farms would be largely self-sufficient in at least some basic staples (grain, potatoes, vegetables, eggs) and, besides fowl, most farms kept pigs to provide meat. In areas where livestock grazing was important, farms had access to milk (cow, ewe or goat) and, if in sufficient supply, possibly butter and cheese. It was also usual to have at least some share in their particular regional alcoholic beverage, whether cider, wine, beer or a spirit such as brandy or *poitín*. In Mediterranean countries, olives were widely grown to press for oil; further north, animal fats were kept for cooking. Fuel needs were met from cutting trees or digging peat from the bog; wool and flax supplied fibre for clothing. Animal feeds were always supplied on the farm, with land set aside for hay, oats and roots.

This situation prevailed across large parts of Europe and in other parts of the temperate world for longer than might be commonly imagined today. While the first decades of the twentieth century marked the early development of motorisation in agriculture, it must be remembered that this was also a period of considerable turmoil marked by the Great War (1914–18) and the Great Depression (1929–33). Thereafter, efforts to increase the dissemination of agricultural innovations were further disrupted by another round of international warfare during 1939–45. It was not surprising, therefore, that by 1945 there was considerable diversity across Europe in levels of productivity, and not only a desire for reconstruction, but a determination to forge an utterly new way of developing agricultural production.

The productivist imperative

Yet, as Mazoyer and Roudart (2006) ask, "How could small and medium-sized farms of a few hectares, practicing diversified production using animal traction and largely self-sufficient, be converted in little more than half a century to large-scale motorised, mechanised, and specialised commodity production?" They suggest that a sequence of gradual transformations were developed one from another, in which successive advances in the mechanical and chemical industries were matched by improvements in the selection of crops and livestock, and in the expansion and specialisation of farms. In other words, this was a *co-evolutionary* process across a number of separate but linked sectors. However, while noting technological advances, Mazoyer and Roudart appear to attach less weight to public policy in effecting such change. In searching for answers to the question posed above, let us look briefly at the experience of the UK.

Box 2.5 provides a picture of pre-war agriculture in Britain, where a mixed farming system made optimal use of on-farm resources. The widespread introduction of modern agricultural technologies after 1945 effectively swept away such pre-war practices and established a framework by which farmers were expected to achieve ever-higher levels of productivity. The means to achieve such modernisation was performed by public policy (that set by national governments and state agencies), which played an indispensable and instrumental role in requiring farmers to adopt new practices involving new technologies.

Box 2.5

Traditional low-input farming, pre-war UK

In Britain, farmers were virtually self-sufficient in farm inputs prior to the Second World War. Such machinery as existed on the farm was made locally or on the farm itself; draft power was provided by horses, raised on the farm and fed from the farm; and seeds were saved from one harvest to the next. Soil fertility was maintained through crop rotations and through rotating crops with livestock, the manure from the animals being used to fertilise the land. Typically, a field might be used to grow wheat, then turnips, then put down to grass interplanted with legumes such as clovers and vetches, and then used as pasture. The planting of legumes was critical to the cycle, since legumes (unlike other plants) have the ability to fix nitrogen from the air and pass it into the soil through their roots.

Crop rotations also helped to combat pests, particularly those pests that are carried by soil organisms. Most pests and crop diseases are specific to individual plants and will not attack other species; a fungus that might devastate potatoes, for example, will not affect beans. If potatoes are grown on the same patch year after year, the fungus is provided with a permanent source of food and may quickly become established as a major pest. But if the potatoes are moved to another field and beans are grown instead, the fungus is not given an opportunity to establish itself.

Source: adapted from Clunies-Ross and Hildyard (1992), 42.

This involved "carrot-and-stick" elements: providing financial incentives (grants and subsidies) to encourage; and penalties, coercion and enforcement in cases of non-compliance. For example, following the 1947 Agriculture Act in the UK, farmers were required to comply with "rules of good husbandry", which was equated with maximising output from the land. Failure to comply rendered farmers liable to supervision orders and even dispossession of their land – between 1947 and 1958, 5000 farmers were placed under supervision orders and 400 were dispossessed (Clunies-Ross and Hildyard 1992).

The logic of agricultural intensification, in which tractors replaced horses, and chemical fertilisers and pesticides displaced rotations, irretrievably altered the entire balance of the mixed family farming system. While grants were available to plough up grazing land, or build new milking parlours, farmers were expected to borrow money to buy machinery and use credit by banks to purchase inputs. Gradually the financial institutions have become an integral part of the new agri-food system, providing a source of credit for new rounds of investment against the collateral of agricultural land. It is in this context that the coercion and enforcement used on farmers to ensure adoption of new practices and technologies now seems especially stark. What was once a largely sufficient unit, forming part of a local or regional mosaic of food producers, has now become encircled by creditors, input suppliers, knowledge providers and those voracious industrial and commercial operators downstream. Figure 2.2, which derives from a study of the agricultural economy of southern Sweden, perfectly illustrates the way in which the farm has become a hub to transform a wide range of inputs into food outputs. That it is required to do so as cheaply as possible is the nature of the agricultural treadmill.

The agricultural treadmill

Essentially, the treadmill appears to offer a means of moving forward, a prospect of progress and improvement. By adopting a new technique or technology, farmers hope to raise yields and increase incomes, possibly even to achieve a windfall profit. But others adopt the technology too, resulting in higher volumes of food entering the market and placing a downward pressure on prices. In other words, farmers' costs go up (purchase of inputs), farm-gate prices go down, and only increased volumes of sales maintain profitability. For example, in the UK, farmers' share of the value of a basket of food items fell from 47 per cent in 1988 to 36 per cent in 2007, a decline of one-fifth in real terms. Yet input costs have continued to rise over this period: this is the nature of the "cost–price squeeze". Those who cannot compete are driven out, leaving their land to be absorbed by others who can now benefit from new economies of scale and lower unit costs. The only way to stay ahead is by adopting the latest technologies, requiring new rounds of borrowing. The metaphor of farmers being required to run up the down escalator has also been employed: only if they can run faster than the escalator can farmers hope to remain solvent or increase their profits. Escalator or treadmill, either offers a vivid illustration of how primary food producers generally have been trapped in a system requiring them to achieve increases in productivity and lower food prices. Box 2.6 provides a summary of the many negative consequences arising from the agricultural treadmill.

It would be a mistake to imagine, however, that the treadmill or cost–price squeeze operates only in the highly developed countries. In this respect, the contemporary agri-food system is truly global, with high-input modernised farming widespread in the South and often

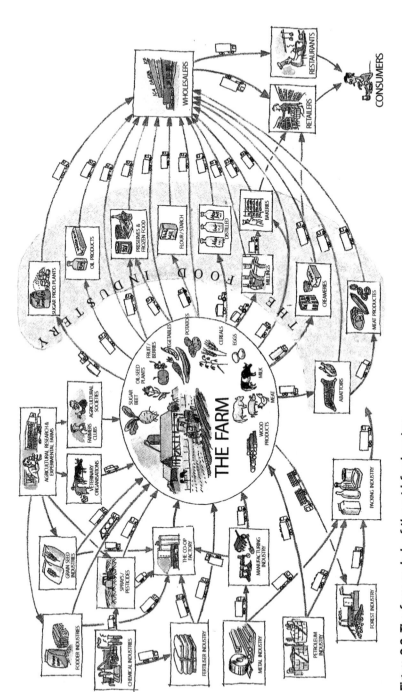

Figure 2.2 The farm as hub of the agri-food system

Source: Buttimer and Stol 2001: 85, reproduced with permission of Cork University Press.

Box 2.6

Features of the agricultural treadmill

- Competition among farmers promotes non-sustainable forms of agriculture (use of pesticides, loss of biodiversity, unsafe foods). The treadmill is contradictory to nature conservation, drinking water provision, landscape conservation and other ecological services.
- The treadmill leads to loss of local knowledge and cultural diversity.
- Not consumers, but input suppliers, food industries and supermarkets, capture the added value from greater efficiency.
- The advantages of the treadmill diminish rapidly as the number of farmers decreases and the homogeneity of the survivors increases. It has a limited life cycle as a policy instrument.
- A global treadmill unfairly confronts farmers with others who are in very different stages of technological development, and have very different access to resources. Although the costs of labour in industrialised countries are many times those in the developing countries, labour productivity in agriculture in the North is still so much greater (and subsidised) that small farmers in the South do not have a chance of developing their agriculture.
- The treadmill leads to short-term adaptations that can be dangerous for long-term global food security and environmental sustainability.

Source: adapted from Röling (2005), 108–9.

initiated with a similar degree of coercion as described for the UK. Pretty (1995), for example, outlines the compulsion used by the Indonesian authorities to force farmers to adopt the new technology packages including high-yielding varieties of seeds and the chemical fertilisers and pesticides so instrumental to their success. This "Green Revolution" technology package, which was promoted in South, South-East and East Asia and in parts of Latin America during the 1960s, 1970s and 1980s, arose largely from plant-breeding efforts that were concentrated in two international agricultural research centres, in Mexico (wheat and maize) and the Philippines (rice).

In these two centres, agricultural scientists concentrated on breeding new hybrid varieties of staple cereals that demonstrated a number of innovative characteristics. They were day-length insensitive, so could be extended to farmers at a wide range of latitudes; they matured quickly, permitting two or even three crops to be grown each year; they were much more nitrogen-responsive than traditional varieties; and were capable of producing more grain at the expense of straw. These modern varieties were then made available to farmers together with high-cost inputs, including inorganic fertilisers, pesticides, machinery, credit and water regulation. These technical innovations were naturally implemented in the best favoured agro-climatic regions, usually with established irrigation systems, and for those classes of farmers with the best expectations of, and means for, realising the potential yield increases. As a result, average cereal yields have more than doubled in thirty years. However, as Box 2.7 describes, decisions made to favour the development of high-yielding varieties of rice, wheat and maize were at the expense of other policy and technology choices.

In summary, then, this section has traced the contributions of farmers and their families in response to efforts to increase productivity. There may be around 450 million farms worldwide producing primary foodstuffs for global, national or local markets as well as for direct consumption by farm families. The overwhelming majority of these farms are small, family-run units, with most occupying less than 2 ha of land, particularly in Asia. In chapter three, the main types of farming system and the way they utilise resources are examined. However, in this next section it is vital to understand the way in which the agri-technologies sector has shaped on-farm production.

Box 2.7

The Green Revolution imperative

Historically, the Green Revolution represented a choice to breed seed varieties that produce high yields under optimum conditions. It was a choice *not* to start by developing seeds better able to withstand drought or pests. It was a choice *not* to concentrate first on improving traditional methods of increasing yields, such as mixed cropping. It was a choice *not* to develop technology that was productive, labour-intensive, and independent of foreign input supply. It was a choice *not* to concentrate on reinforcing the balanced, traditional diets of grain plus legumes.

Moreover, in light of these "paths not taken", we must ask ourselves: in our eagerness to embrace the new, in our rush to extend the scope of human knowledge and control, do we forget to work on *applying* the collected wisdom already handed down to us? Has our fascination with science prevented us from tackling the incomparably more difficult problems of social organisation and the agricultural practices of real farmers? For the majority who are hungry, "miracle" seeds are meaningless without control over land, water, tools, storage and marketing.

Source: adapted from Lappé and Collins (1982), 114.

5. Agri-technologies

Scientific and technological advances that have taken place away from the farm have arguably exerted a greater influence over the kinds of food produced – and how – than decisions made by farmers over the past sixty years. The development and application of mechanical, chemical and biological engineering in response to the challenge of increasing productivity have fundamentally altered the relationship between agriculture and the environment, as well as definitively ending the autonomy of farmers. This section briefly reviews some of the key developments in the agri-technologies sector that have had such a bearing on the nature of farming. In chapter three, more detailed attention is given to some of the environmental consequences arising from the application of these technologies.

Mechanisation

Mazoyer and Roudart (2006) examine in detail the development of what they term "motomechanisation" in agriculture. Beginning in the inter-war period in the temperate regions of European settlement and then into Europe, they emphasise how the deployment of this new oil-based technology only really took over from animal traction after 1945. Initially applied to grain and oil seed crops, which were widely grown in these regions and on a large scale suitable for mechanisation, efforts were directed to tractors and attachments for ploughing and sowing, and to (combined) harvester–threshers. Motomechanisation only later spread to the harvesting of row crops (potatoes, beet), and then to other more specialised field tasks such as harvesting fodder, vegetables and fruit crops, including grapes. It was also developed for and applied to the milking of dairy cows, replacing hand milking that restricted the size of a herd to about a dozen cows. Today, many of the commercial dairy farms, generally with more than 100 cows, feature herringbone or rotary sheds that have speeded up the milking process significantly, with the latest development comprising robotic milking systems in which each cow is trained to present itself twice over a twenty-four-hour period. Such technology is designed to overcome limits on the number of cows that can comprise a dairy herd, with some highly industrialised milk farms in the USA now managing herds in excess of several thousand milking animals. Box 2.8 sets out the stages of motomechanisation according to Mazoyer and Roudart, and Figure 2.3 illustrates the productivity achievements as a result.

Box 2.8

Stages in the process of motomechanisation

Motomechanisation I: Sees the replacement of draught animals (and a few rare steam tractors) by tractors with low-power internal combustion engines (10–30 horsepower, hp). Usually attached to pre-existing, animal-drawn equipment (ploughs, reapers) and transport (carts, wagons). This stage, which began in the inter-war period, spread rapidly from the end of the 1940s to farms of more than 15 hectares (ha) and capable of buying and making use of them. Though not very powerful, they were faster than animals and made it possible to increase the area per worker from 10 to 20 or 30 hectares, typical of large-scale farming at that time.

Motomechanisation II: Emergence of medium-power tractors (30–50 hp) generally fitted with lifting mechanisms enabling them to carry, rather than just pull, attachments, and with power take-off to drive other machines. This generation of tractors made it possible to draw two ploughshares (thus doubling productivity) as well as harrows, rollers and spreaders. New equipment was designed and manufactured for these tractors: grain harvester–threshers, potato and beet harvesters, and so on. In Europe, this stage marks the end of the 1950s and 1960s. Compared with stage I, it made possible the doubling of farming area per worker to 50 ha.

Motomechanisation III: This stage sees tractors of 50–70 hp, capable of pulling three-furrow ploughs and implements up to 6 m long. It also marks the appearance of large, self-propelled combined harvester–threshers with frontally arranged cutting machinery of greater width than those of tractor-drawn machines. This stage developed at the end of the 1960s, and made possible an increase in the area per worker in large-scale farming to 70–80 ha.

Motomechanisation IV: Tractors of 80–120 hp capable of pulling four-furrow ploughs. Attached equipment was itself becoming increasingly sophisticated in performing simultaneously complementary field tasks. Combined harvester–threshers with a width of cut of up to 6 m appeared. This stage spread in Europe during the 1970s and 1980s, and led to an increase in the number of hectares per worker to more than 100.

Motomechanisation V: The current stage in large-scale intensive farming in Europe (though possibly being superseded in the USA?), this stage features the development of four-wheel-drive tractors of more than 120 hp and the use of associated equipment that makes it possible to prepare the soil and sow seeds in only one run. The area per worker can now exceed 200 ha.

Note that, while representing an enormous saving in human labour, the fixed capital invested per worker increases from around US$50 in stage I to $500,000.

Source: adapted from Mazoyer and Roudart (2006), 382–3.

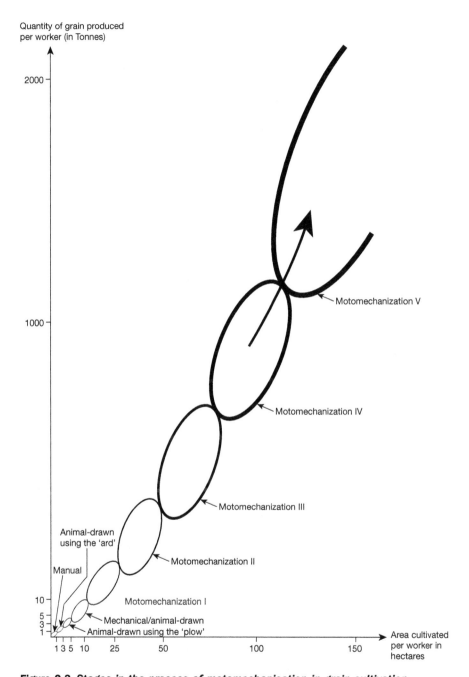

Figure 2.3 *Stages in the process of motomechanisation in grain cultivation*

Source: Mazoyer and Roudart 2006, p.384, reproduced with permission of Earthscan Ltd, www.earthscan.co.uk.

Mechanical technologies have consequently become increasingly powerful and sophisticated, but it is worth briefly recalling the early efforts. In 1917, according to Feehan (2003), a Belfast man, Harry Ferguson, was approached by the Department of Agriculture for Ireland with the challenge of improving the quality and efficiency of mechanised ploughing. The essentials of the new tractor were in place by the early 1930s, and the first Ferguson tractors manufactured by David Brown of Huddersfield sold for around £244. In 1938, Ferguson teamed up with Henry Ford, and in 1953 joined forces with Massey-Harris of Canada.

Today, the Massey-Ferguson brand forms part of a large consolidated global agricultural machinery corporation called AGCO, which includes a number of other long-established manufacturers including Fendt of Germany and Valtra of Finland. Yet, despite its size, with over 11,000 employees and sales of nearly US$3 billion per year in the early 2000s, AGCO ranks only third in the global list of agri-machinery companies. Top is John Deere, a US company with 50,000 employees worldwide and total net sales and revenues of US$28 billion in fiscal year 2008, helped no doubt by an increase in sales of agricultural equipment of 37 per cent over the previous year. The second-ranked company is CNH, which was created through the merger of New Holland and the Case Corporation, and which forms part of the FIAT Group. It has 28,000 employees in thirty-nine manufacturing facilities throughout Europe, the Americas, China, India and Uzbekistan.

The identification of the top three agricultural machinery companies gives some indication of the degree to which the sector has undergone significant global consolidation (through merger and acquisition) and demonstrates a high degree of market concentration, not only in common with most other branches of the inputs supply element, but in keeping with the agri-food system in general, as we shall see. Concerns with the emergence of monopoly in the farm machinery industry have been a long-standing issue: for example, in the first half of the twentieth century in the USA, there were six government investigations of the industry, and most of these were the result of complaints made by farmers about the proportionately greater decline in farm product prices than machinery prices at times of economic depression. By 1986, the top four companies accounted for 80 per cent of tractor sales in the USA, and it is likely that today the concentration ratio for the top four is greater still. However, after tractors and combine harvesters, there appears to be a larger number of smaller companies producing more specialised equipment for specific purposes in regional markets. Such companies remain vulnerable to take-over and acquisition by the large conglomerates, however.

Chemical fertilisers

If the nineteenth century saw the development and application of mechanical technologies for improving the productivity of labour in agriculture, it also witnessed something of a golden age of soil chemistry. Throughout this century, scientific research gradually unlocked the contribution of different chemical elements. By the 1860s, according to Tudge (2003), it was known that plants require significant amounts of nitrogen, phosphorus, potassium and sulfur, as well as magnesium and iron; the only uncertainty was in what quantities. Trace minerals such as boron, zinc and manganese are required in only tiny amounts, but if entirely absent from the soil can result in deficiency and disease in growing crops. Tudge illustrates this by recounting a situation in the Terai zone of Nepal, where chickpeas were dropping their flowers and failing to set seed. Eventually, it was recognised that boron was the deficient mineral: adding it to the soil at a rate of just 1 kg/ha almost doubled yields in some areas.

Yet if there is one element that has proven the single most important limiting factor in soils, it is nitrogen. As Box 2.5 outlined, farmers in Britain – as elsewhere – practised rotations where legumes such as clover (or beans) fixed nitrogen in the soil via nodules on their roots, which not only enabled them to grow, but left behind a surplus that could be used by a following cereal crop. This phenomenon helps to explain the strong grains–legume association in traditional farming systems in many different parts of the world, and also demonstrates the resulting combination that composed the dietary staple: tortillas (made with maize flour) and beans in much of Mexico and Central America; chapattis (with wheat flour) and/or rice with pulses (daal) in South Asia; pigeon peas with millet or sorghum in parts of Africa, and so on. Although legumes have fixed enormous quantities of atmospheric nitrogen in soils for thousands of years, it is with the development of the technology to synthesise nitrogen in an industrial form that the history of soil fertility was changed.

Though initially designed as an aid to Germany's 1914 war effort in making the explosive TNT, the invention of Fritz Haber and Carl Bosch has had a dramatic bearing on the course of human history. If soil fertility had once been a matter for farmers to organise through field rotations, the gathering and application of animal manure or seaweed, or even the occasional use of more exotic amendments such as Peruvian guano, it was now a product of the factory. It has been estimated that the annual turnover of nitrogen in the world's crops (that is, the amount taken up and then removed on harvesting food crops) is 175 million tonnes (mt), and almost half of this is provided by synthetic nitrogen produced by the Haber–Bosch process. Of the remainder, 30 mt is supplied by biological fixation, 30 mt from mineral sources, and just 15 mt from animal manure (Tudge 2003). The contribution of artificial fertilisers, together with development of fertiliser-responsive, high-yielding varieties of seed, have led some to argue that up to half of the present population of the world today owes its existence to the availability of synthetic nitrogen for food production, and our dependence upon it is truly irreplaceable (Smil 2000). Yet the dramatic increase in the use of nitrogen fertilisers has raised concerns that farmers are caught on a chemical treadmill, applying more and more fertiliser as a way of staying ahead of the cost price squeeze, though with serious consequences for soils and the wider environment, an issue addressed in chapter three.

Pesticides

If, as Tudge suggests, the development of synthetic nitrogen was enough to bring the entire food supply chain within the purlieus of industry, particularly the chemical industry, this control was further tightened by the development of pesticides. An outbreak of downy mildew in French vineyards during the 1880s was resolved with applications of copper sulfate mixed with lime. By the 1920s, "Chemists were supplying vicious and corrosive inorganic agents to kill everything" (Tudge 2003). During the 1930s, however, the development of hormone-based herbicides, which mimic chemical processes within the weed, enabled a more selective and targeted approach to eliminating competition for crop nutrients. By the end of the decade, DDT had been synthesised and had become widely used as an insecticide, effectively controlling typhus and malaria without causing a significant impact on human health. After 1945, other organochlorine pesticides were developed, and this seemed to herald an age of complete human mastery of the biosphere: human ingenuity would dominate and control nature for our collective benefit.

Yet the 1960s began to reveal the consequences of this rationale: the accumulation of DDT and other pesticides in natural **food chains** so that top predators (such as eagles) showed high concentrations of chemicals originally sprayed to eliminate mosquitoes. This caused them to lay eggs with thin shells that broke during incubation, resulting in a collapse of these important, charismatic species. Rachel Carson's (1962) book *Silent Spring* drew attention to the excessive application of such chemicals, and has proven to be one of the most important texts of the environmental movement. While environmental regulations have curtailed sales of certain pesticides in the North, high rates of sales continue to be achieved in the South, with the fruit and vegetable export sector recording accelerated rates of pesticide use (Dinham 2005).

As a result of pressure on sales, tighter regulations, and the costs of research and development, the large agri-chemical manufacturers have been engaged in a process of merger and consolidation. This has resulted in the creation of six companies effectively controlling around 80 per cent of the world's agrochemical market (Dinham 2005). This process was not simply to build market share in pesticides, however, but to extend their interests into seeds and veterinary products, principally pharmaceuticals. Their rebranding as "life science" companies marks the rise of genetic modification in which the link between seeds and proprietary pesticides has been the characteristic feature of the first stage in the era of agricultural biotechnology.

Seeds

It is with regard to the control over seeds of food crops where the displacement of farmers by global transnational corporations is arguably best illustrated. For it should be acknowledged that the stock of genetic resources represented by the enormous diversity within different food crops is principally the result of hundreds of generations of farmers engaged in selection and deliberate cross-fertilisation to produce plants with traits most suited to local growing conditions and desired eating qualities. This material might be regarded as a collective inheritance, a form of common property, to be saved, shared and exchanged between farmers and other growers of food. With the emergence of commercial hybridisation, however, this balance began to change.

Hybridisation is effectively the crossing of distinct but not unrelated strains of the same species to produce a hybrid variety that is usually more vigorous than either of the parent plants. Although practised by farmers, it was with the onset of mechanisation and chemical fertilisation that commercial interests began to develop new uniform varieties. Plant breeding involved building responsiveness to fertiliser, for example, so that soil nutrients would be converted into more grain or tubers than more lush and abundant vegetation. As grain panicles grew heavier, however, it was necessary to breed shorter stalks to prevent the toppling over ("lodging") of the plant. Besides productivity (yield), it was also necessary to establish uniformity in the growing crop, ensuring that plants grew to the same height (for ease of mechanical harvesting) and matured at precisely the same time, and perhaps incorporated other characteristics such as resistance to disease (e.g. wheat rust or potato blight). The breeding of livestock was also carried out along similar lines, with the goal of improving productivity measured in terms of the conversion ratio of feed to weight gain, milk yield, or numbers of offspring per year, as well as selecting for ease of mechanisation, as in the case of creating uniform teats on animals bred for battery milking.

The development of hybrid seeds has certainly increased the volume of food output, associated as they are with the entire panoply of industrial agricultural inputs. However, they are also associated with an accelerating loss of genetic diversity, largely as farmers surrendered their traditional varieties in place of the new "miracle" seeds that promised higher yields. These seeds have proven an effective way of encouraging farmers in the South to climb onto the agricultural treadmill: requiring them to purchase fertilisers, pesticides and machinery in order to maximise productivity, even displacing food crops for family consumption to increase the area under commercial cultivation. Yet hybrid seeds significantly extend the technological dependence of farmers on private companies that have successfully enclosed the commons of publicly held genetic resources with patented control over the means of growing food. This process has been significantly extended with the development of genetic engineering.

This chapter demonstrates the degree to which resources and technologies previously in the hands of farmers and local artisans were taken from their control and transformed into industrial inputs for their purchase. This is a process known as **appropriationism**: each aspect of agricultural activity (soil fertility, ploughing, seed selection) becomes the basis of an autonomous sector for innovation, technological control and the accumulation of value in places far distant from farmers. Though some may argue that this is the price to be paid for "progress" to achieve food surpluses, it should be remembered that most of the hungry and the food insecure today are (and were) farmers impoverished by this model of agricultural development. The loss of accumulated local knowledge of farmers, as well as their genetic resources, leaves us vulnerable to all manner of ecological perturbations. Yet it may not be too late to attempt to restore some balance between primary producers and their upstream neighbours.

6. Food trading, processing and manufacturing

This component of the agri-food system involves the procurement of primary products and then undertakes their transformation into a wide range of final foods (for purchase by consumers). It is represented by some of the most iconic global brands that, to some extent, personify the consumer-oriented and commercially successful nature of the food system. Many of the leading manufacturers – Nestlé, Kellogg, Cadbury, Coca-Cola – were founded back in the late nineteenth century and have remained dominant players in the market by dint of their **horizontal consolidation**. This refers to a process whereby a company merges with, or acquires, another within the same sector, thus increasing its share of that particular market. As noted earlier, for manufacturers seeking to build a presence in new markets overseas, taking a stake in, or outright purchase of, a local company is a common method to establish a production platform and distribution network. It also, of course, can help to extend product range outside the traditional core brands of the company. In addition, some companies have engaged in **vertical integration** through the acquisition of existing businesses up- and/or downstream of its existing operations. This has been a feature, for example, of some grain-trading companies that not only have built considerable capacity in their core business of purchase, storage and transportation of cereals and oil seeds, but also have sought to add value to that material by processing it into animal feeds, and then to add further value still by controlling the conversion of that feed into meat.

Exercising corporate reach

The emergence of food traders/processors such as Cargill and ConAgra during the twentieth century is indicative of the deteriorating bargaining position of primary production within the overall food system. This is illustrated by the widespread use of contract production, whereby traders and processors (as well as retailers, as we shall see) specify the methods of production, timing and harvest dates, the uniform characteristics of size and appearance – particularly for fresh fruit and vegetables, which must be blemish-free – and the price that will be paid. Contract production has enabled some diversified companies effectively to surround the producer, being simultaneously upstream in the provision of inputs and downstream in taking delivery of the output of primary production. Perhaps the best illustration of this process is the US meat industry.

Describing the transformation of hog (pig) production in the United States, Page (1997) highlights the advances in animal science that enabled producers to concentrate larger groups of animals at a single site through advanced confinement practices. The development of a wide array of inputs testifies to the way in which specialised agri-technology industries develop in response to changes in production practices that are driven from the food-processing sector. Thus inputs have been developed such as boar semen (with the development of artificial insemination techniques), synthetic hormones and antibiotics (for increased growth), manufactured feed, and the installation of specialised buildings together with mechanical systems for heating, cooling, feed delivery, waste removal and storage (recalling the parallel provision of industrial inputs in Figure 2.1). These inputs represent costs to be paid for by the primary producers – and, in the case of investments in new animal housing facilities, require farmers to undertake significant borrowing as a required entry cost to participating in the sector – and while carrying the burden of risk (e.g. outbreaks of disease), the rewards are determined by the external contractor. Page explains:

> With the entry of ConAgra and Cargill, much of the pork packing industry is now housed within large and diversified enterprises that operate across a variety of commodity chains and around the world. In the US, these firms are involved in nearly every stage of pork production, including inputs to grain farming, grain transport and processing, livestock feed manufacturing, meat packing [ie. slaughter and disassembly of the carcasses], and processed foods production. Neither firm has a significant involvement in hog production, however . . . (which) continues to be dominated by mid-sized farm enterprises using family labour augmented by some hired labour . . . Overall, the diffusion of capital intensive techniques has led to dramatic reductions in the number of farms raising hogs and a concomitant increase in the average number of hogs produced per farm.
>
> (Page 1997: 139–40)

Corporations such as ConAgra and Cargill, which built their early success on grain trading, are now highly diversified operations with global interests across a range of sectors and enjoying, according to many economists, a disproportionate influence in agri-food markets. One commonly used measure to indicate the degree of horizontal consolidation, or concentration, in a market is the four-firm ratio (CR4). This simply comprises the aggregate market shares of the four largest firms operating in a particular sector, and is expressed as a percentage. In general, economists who use CR4 believe that as it rises above 40 per cent, the market's competitiveness begins to decline. A market with a high CR4 might be considered as an effective oligopoly, where a few companies have a greater ability to control prices and to conceal market information to the detriment of other actors (producers, smaller competitors, consumers). Indeed, circumstances where a small group of companies enjoy a high degree of control can lead to suspicions of a price-fixing cartel engaged in anti-competitive practices that contravene legal–financial norms. Several high-profile cases in recent years have revealed the involvement of large corporations engaged in conspiracies to fix prices in agri-food markets in developing countries. It is no wonder that Dwayne Andreas, one-time Chairman of one of these indicted companies, Archer Daniels Midland, was reported to have said to his son, Michael, another former executive of ADM, "The competitor is our friend and the customer is our enemy." Examples of CR4 for different sectors are shown in Box 2.9 and the reoccurrence of some named corporations should be noted.

The food-processing and manufacturing component consequently demonstrates the inherent logic of the system, which is to take low-value, undifferentiated primary commodities and to add value through a succession of industrial processing activities. Initially founded by building upon long-established food transformation techniques with new patented innovations, for example Nestlé's dried and evaporated milk, and J.H. Kellogg's cornflakes, this food industry created a range of durable, convenient and easily prepared foods. Yet in its drive to enhance profitability, innovation was often spurred by the apparent desire of the industry to reduce dependence upon specific raw materials, either through the search for interchangeable inputs (replacing barley for wheat, for example, if its price was lower), or even to eliminate it entirely, replacing it with industrial substitutes wherever possible. Possibly the best example here is the displacement of natural sugar (both cane and beet) by artificial sweeteners such as saccharine, aspartame and sucralose; there are also a host of natural sugar substitutes, of which high-fructose corn syrup is the most widely used

---

Box 2.9

Concentration ratios for selected US food sectors

(CR4 unless otherwise stated.)

- Grain handling: 60 per cent (2002) [in Mississippi Gulf ports that handle 70 per cent of US maize exports, the four largest firms own 98 per cent of all storage capacity].
- Grain milling: 63 per cent (2003) [Cargill, ADM, ConAgra, Cereal Food Producers].
- Soybean crushing: 80 per cent [CR3 = 71 per cent: ADM, Bunge, Cargill].
- Beef slaughter and processing: 84 per cent (2003) [Tyson, Cargill, Swift, National Beef Packing].
- Pig slaughter and processing: 66 per cent (est. 2006) [Smithfield, Tyson, Swift, Cargill] up from 37 per cent in 1987.
- Broilers: 59 per cent [Pilgrim's Pride, Tyson, Perdue, Sanderson].
- Turkeys: 55 per cent [Butterball, Hormel, Cargill, Sara Lee].
- Inputs: US market for maize seed: CR2 = 58 per cent [Monsanto, Dupont]. Brazil: pesticides market for soy, corn, beans: CR4 between 85 and 100 per cent depending upon crop. Similar CRs can be found in other countries.

Sources: adapted from Harkin (2004); Hendrickson and Heffernan (2007); Weiss (2007).

---

in the food-manufacturing industry. This process is referred to as **substitutionism**, and it has effectively marked a very clear separation of interests between agriculture and primary processing activities, and those of final foods manufacturing.

The era of innovation

Yet, whereas the late nineteenth and first half of the twentieth century was marked by "Fordist-type" industrialisation of food processing technologies (essentially the scaling-up of production, but of a limited range of products), the second half of the twentieth century witnessed an apparent shift toward a more consumer-oriented approach. In place of a single product brand (available, at best, in either large or small size), food industries were required to embark upon a strategic diversification, developing, testing and marketing a wide range of new products, if they wished to remain major players within a rapidly changing food market. Market analysts speak of "segmentation" to describe an increasing heterogeneity within the market based on gender, age, income and lifestyle characteristics, each with its own food preferences. A multi-product strategy was consequently designed to enable a manufacturer to tick as many of these boxes as it could, and one of the easiest ways of doing this was simply for it to take over another company with a product range that extended the existing range of brands.

One of the critical factors that help to explain this shift to product diversification has been the rise of the corporate retailers, the focus of the next section of this chapter. Their rapid expansion from the 1970s onward has definitively altered the balance of power in the food system, partly as a consequence of their greater proximity to the consumer, and the way they have come to represent consumer interests in their dealings with both the public and food manufacturer–suppliers. Clearly there is a close and dynamic interconnection between the kinds of food products being developed by manufacturers and the ways in which these are presented in store. The shift to prepared foods, for example, presupposes the ability of manufacturers to source and utilise a wider range of ingredients and actively to appropriate many of the tasks previously performed in domestic kitchens. With the development of "heat-and-eat" ready meals, which were effectively to take off once the more thoroughly innovative chill-chain technologies (including packaging) were able to replace the antediluvian canned meal, the appropriation of cooking seemed complete, the skill of manufacturers' product developers was apparent, and retailers' refrigerated cabinets would supply an evening meal at the end of a customer's busy working day.

"With the shift from single product transformation to prepared meals and ever more sophisticated snacks", argues Wilkinson (2002), the industrial food factory becomes an immense kitchen-*cum*-laboratory, a source of great internal innovation with complex demands on equipment and ingredients suppliers. This is particularly true with regard to the development of snacks and beverages, especially soft drinks, where there has been such significant growth potential and where some of the largest and most powerful food businesses can be found. The source of their wealth has been in their ability to create markets for products, such as carbonated drinks or confectionary, of limited or no nutritional value, and to make them ubiquitous and, through advertising, desirable.

For example, a company such as Nestlé, the world's largest food and beverage manufacturer with a turnover in 2007 of US$95 billion and a 3.4 per cent share of the global market for sales of packaged food, now has interests well beyond its core competence and raw material inputs traditionally associated with milk products, coffee and confectionary. Today, its corporate structure contains divisions responsible for bottled water (including Perrier), breakfast cereals, dairy and ice cream (including the Haagen-Dazs brand), pet foods, prepared foods (such as "Lean Cuisine") and baby foods. Nestlé, it should be noted, is a manufacturer of infant formula feeds and has been subject to a long-standing campaign that believes it to violate agreed codes of conduct with regard to the marketing and promotion of formula feeds in countries of the South, to the detriment of breastfeeding (Baby Milk Action, www.babymilkaction.org). Despite that, Nestlé describes itself not as a food, but as a "nutrition, health and wellness" company.

Yet, while drawing attention to the size and undoubted influence of the largest agri-food corporations, it is vital to understand that this component of the system comprises millions of small and medium-sized enterprises (SMEs) around the world. Within the European Union (EU-27) in 2005, there were approximately 310,000 food and beverage manufacturing enterprises providing employment to 4.7 million people and generating turnover of €850 billion. France and Italy have the largest number of such enterprises by some distance, though Germany has the largest workforce, followed by France and the UK. This gives some general indication of the number of small specialist artisan food producers characteristic of France, Italy and other countries in Southern Europe, compared with larger-scale manufacturing symptomatic of Northern Europe.

7. Food retail and the food service industry

This section explores the final box in the global agri-food system (Figure 2.1), comprising food retailing and food service. Although they form a paired endpoint of the food chain, they are treated separately. Food retail refers to the sale of fresh and packaged food and drink for consumption at home, while food service concerns the sale of prepared foods and beverages that take place outside the home. During the past twenty to thirty years, food retailing has been utterly transformed by the rise of giant retail corporations, and the food service sector, too, has grown rapidly over this same time scale. Market leaders in both sectors are global multi-billion-dollar operations, and those that have remained largely regional or national in scope are coming under increasing pressure to retain their market share and their independence.

Food retailing

Beginning with retailing we should, of course, acknowledge that prior to the rise of the giant corporate retailers, with their hypermarkets, supermarkets and other outlets, people frequented markets – the periodic, open-air markets found in squares and streets across much of Europe, as well as the covered municipal markets created as a public health measure from the mid-nineteenth century onwards. There were also independent specialist shops (the butcher, the baker, the fishmonger), mobile street traders (milkmen) and, from the late nineteenth century, grocers such as Thomas Lipton with shops selling dried goods (tea, sugar), cured meats, dairy products and other provisions. The supermarket emerged in the United States in the early part of the twentieth century, and it was estimated that by 1940 it accounted for 40 per cent of all food shopping in the USA. Although the format of self-service shopping was adopted more slowly in the UK, supermarkets grew rapidly in number through the 1960s and numbered 3400 by 1969. Their growth seems in part to reflect a more egalitarian spirit: irrespective of class, gender, age or ethnicity, customers could buy their food needs in the quantities they required without having to deal with counter service staff. Since then, of course, the supermarket retail chains have become the dominant suppliers of food and drink throughout the North and are making increasing headway into countries of the South.

At this stage in the agri-food system, we come to a point where the principal gravitational force is no longer supply-led, but is driven by the pull of demand. Whereas food manufacturing is based on the transformation of primary materials into final foods, retailing is predicated on maximising the sale of goods, preferably to a widening group of consumers. This means that, to some degree, retailers are Janus-faced:

● they engage in a dialogue with consumers not simply to understand their needs, but to construct and continuously reshape them in such a way that there is a close correlation between these "needs and wants" and their provision of retail services;

● on the other hand, in their dealings with food manufacturers and suppliers they speak on behalf of consumers, pressing for new product lines, higher standards of food safety and traceability and, above all, reductions in price.

What this focus on price has effectively done is to reconfigure, again, the balance of power within the agri-food system, with food manufacturers and other suppliers now becoming "price takers" rather than price setters. In their drive to expand their market share of total retail sales of food, the competition between corporate retailers has led to a continuous stream of commercial innovations and predatory practices, which have put greater pressure on upstream suppliers. It has also extended the unending quest for optimal economies of scale, with a tendency for all production and retail operations to grow bigger to drive down unit costs. This search for economies of scale has the following results:

● Bigger retail stores able to contain a greater range of items (or stock-keeping units, SKUs, in the jargon of the sector: a typical Walmart has 60,000 SKUs) including a greater volume of non-food items. As clothes, household and electrical goods have a higher profit margin than food items, they effectively serve to subsidise food sales.
● Bigger stores are designed to draw more customers, and this invariably means moving them out of city centres to suburban or green field locations where there is abundant space for car parking. While this is designed to enhance customer convenience, it has enormous implications for urban planning policy and the development of road infrastructure. It also takes away potential footfall from independent city centre retailers.
● More customers and greater turnover generally mean the retailer is selling more of a manufacturer's product. This gives the retailer the upper hand in negotiations on a range of supply issues. Walmart, for example, has a fearsome reputation for its ability to drive down the unit price paid to suppliers, with the volume of unit sales it achieves compensating for the small margins.
● With regard to delivery, all the major retailers have moved to a "just-in-time" supply chain-management system that reduces to an absolute minimum the length of time they hold stock. Retailers have been at the cutting edge of developing information technology that integrates barcode scanning at point of sale to computer-controlled stock management. This is tied into road transport logistics so that a

seamless movement of stock can move from the suppliers into the retailers' own regional distribution centres and then on to their individual store (see chapter five).

- In the case of fresh produce, for example, suppliers agree months beforehand to deliver contracted volumes of produce of designated quality on a specified date. Because of the risks and uncertainty in growing food (pest- or weather-induced crop losses; difficulties in harvesting the crop on time), producers run the risk of failing to meet the specified delivery window, or might fail to meet the "quality" criteria for the product, of size and weight or absence of blemish. Such failings on the part of a producer expose them to the risk of losing all future contracts with that retailer to a competitor.

- Unsurprisingly, with retailers continuously searching for economies of scale, pressures are quickly transmitted to manufacturers to "go large" in an effort to drive down costs. Scaling up production requires building larger facilities and the speeding up of production lines; increased investment in fixed capital (plant and machinery) is designed to reduce labour costs either through a reduction in the numbers employed, or through deskilling and increased use of cheaper migrant labour (the food industry, both in the fields and in the factories, has a long association with the recruitment of undocumented, i.e. illegal, and therefore cheap, migrant workers).

- There are many other issues that reveal this unequal balance of power: slotting fees, for example, that manufacturers pay in order to have the right to have their product on the shelf; or their heavy discounting of the product so that retailers can run special offers ("Buy One, Get One Free") in order to bolster customer loyalty; or their development of "own label" products that generate higher margins, undercut established brands and enhance corporate image; or even the strict hygiene standards which they impose on suppliers that are so much more demanding (e.g. in terms of microbiological cell counts) than those required by official public health standards. This extraordinarily powerful role that retailers have come to occupy within the food system can be represented as the pinch point within an egg timer or hourglass, as shown in Figure 2.4.

All these issues are designed not only to improve financial performance and profitability, but also to enhance the standing of retailers in the eyes of the public. The leading corporate retailers have so effectively positioned themselves to speak on behalf of their customers that they have almost become champions of consumer rights. Indeed, some

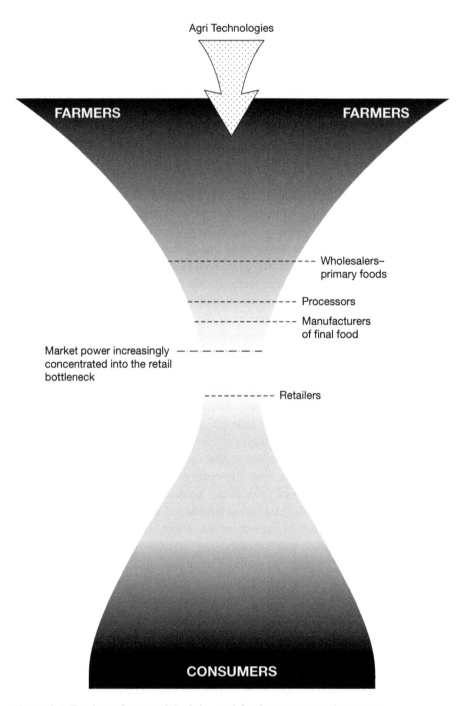

Figure 2.4 *The hourglass model of the agri-food system: retailer power*

retailers have astutely read the prevailing attitudes of their core customer base and taken action to reflect their attitude: the decision of Marks & Spencer to eliminate genetically modified ingredients from its line of foods demonstrates its effort to bolster its position as catering to a higher-price, higher-quality sector. Responding to criticisms of its operations, Walmart went on the offensive, making a series of announcements of its commitment to sustainability, including the promotion of fresh organic produce, and most of its competitors subsequently followed suit. However, it is in relation to price where corporate retailers have truly excelled at winning customers, and as the most successful have grown, they have effectively eliminated much of the small independent sector.

Table 2.2 provides a list of the top five grocery retailers ranked by turnover. It should be noted that Walmart, the world's largest retailer, reports a volume of sales that exceeds the aggregate turnover of the next four companies on the list. Having built their respective national markets, most of these companies have more recently turned their attention to expanding their operations overseas (with the exception of Kroger amongst the top five companies). The rationale for this is straightforward: the emerging markets of the South offer much higher rates of growth in sales than is possible in the highly competitive home markets, where each percentage increase in market share must be won from rival companies. In contrast, East and South-East Asia and Latin America have proven extremely lucrative regions for foreign direct investment by corporate retailers.

Table 2.2 *IGD grocery turnover league, 2007*

Turnover 2006–07 (rank)	Retailer	Country of origin	Net sales 2006–07 (US$ billion)	Percentage change 2005/06–2006/07	Turnover 2008 (local currency, billion)
1	Walmart	USA	348.65	11.7	$374.5
2	Carrefour	France	97.86	6.6	€87
3	Tesco	UK	78.59	10.9	£51.8
4	Metro Group	Germany	75.23	7.5	€68
5	Kroger	USA	66.12	6.2	$76

Source: compiled from IGD (2007); companies' annual reports (2008 figures).

There are a number of factors that help to explain this expansion, but amongst them are the far-reaching socio-economic changes that have spread through many countries of these regions. Rapid urbanisation, generally rising incomes, and associated growth in manufacturing and service industries employing both men and women have generated unprecedented levels of disposable income. Lifestyle changes, the adoption of more western-style consumption patterns and the increasing prevalence of domestic refrigerators have also caused a shift in food purchase behaviour, away from traditional outlets in favour of one-stop shopping. According to Reardon *et al.* (2005), Latin America led the way in the growth of the supermarket sector, which had a 10–20 per cent share of food retailing in the 1980s but achieved 50–60 per cent by 2000. Indeed, fully 75 per cent of food sales in Brazil now occur through supermarkets. And while Latin America may have led the way in this retail revolution, the current wave of supermarket expansion is taking place in China, India and other countries in Asia and, to a lesser extent, in Southern Africa. Here is where traditional retailers are experiencing the greatest pressure.

While these early supermarkets were invariably domestic businesses funded by national capital, the rapid growth of this sector is entirely attributable to market penetration by international retailers. Although initially entering new markets in joint ventures with local partners, these international retailers often used this collaboration as an opportunity to understand better the characteristics of the domestic market before either going it alone or buying out the partner. Consequently, many emerging economies in the South have become highly competitive territories, in which significantly less powerful domestic retailers are losing market share to foreign supermarket chains, which are in turn competing strenuously with each other for market dominance. While the short-term beneficiaries might be middle-class urban consumers enjoying lower prices, the wider consequences for building equitable and sustainable domestic food sovereignty are less clear.

Besides their international expansion, a second key aspect is to understand how retailers have developed a number of different formats in which they can cater to the various and changing needs of food shoppers: from the giant hypermarket to the small convenience store on the high street or the filling station forecourt. Their ability to be always available with low prices is continuing to result in the decline of independent retailers, who either go out of business or join one of the "symbol" groups (e.g. Spar) as franchise operators. The entry of the hard-discount multiples into the food retail market has further raised the stakes. With the financial crisis of 2009, the major discount chains have recorded high rates of growth as customers dispense with loyalty to their usual retailers in search of even cheaper prices. Though lacking many of the known brands within particular lines, and often carrying less than 1000 SKUs, customers seem to be attracted by rock-bottom prices. In 2007, the German discount retailers Aldi and Lidl together accounted for 58 per cent of the European discount market with combined sales of €64 billion. Box 2.10 provides a brief synthesis of some of the key features of the top five retailers.

Box 2.10

Some key features of the leading retailers

Walmart is the largest private company in the world, not just in grocery retailing. Sam Walton opened his first store (as Wal-Mart) in 1962 in Arkansas, based on the principle of selling everyday stuff cheaper than anyone else. This created "a culture of looking for every penny of cost savings that could be wrung out of designs, packaging, labour, materials, transportation, even the stocking of stores", a way of doing business that has become known as "the Wal-Mart effect" (Fishman 2006). This term is used to describe the process where Walmart – or any other retailer – opens a new "big-box" store in a suburban location and its lower prices immediately reshape local shopping habits and begin the demise of small retail competition. Besides its Walmart stores, it operates supercentres that have a floor area of up to 18,000 m², neighbourhood markets and Sam's Club discount stores. Today, Walmart operates 4100 stores in the USA and a further 3100 stores in thirteen other countries, including the UK (through its subsidiary Asda) and Mexico (where Walmex now accounts for around one-third of all retail food sales).

Carrefour is a French company and currently the largest European retailer, with turnover of €82.15 billion in 2008. While almost 46 per cent of its business is done in France, that leaves nearly 55 per cent performed elsewhere, largely in Europe (especially Spain), with 10 per cent in Latin America and 7 per cent in Asia. While Carrefour operates a range of different formats, it is revealing that of its 1750 supermarkets and over 1200 hypermarkets, almost 90 per cent of the former are located in Europe (including France), while nearly half of its hypermarkets are in Asia and the Americas (largely Argentina and Brazil). It also possesses the format of "hard discount stores", with 4800 units, 40 per cent of these located in Spain – which the company describes as its "hard discount territory and testing ground" – and over 900 stores in China, Argentina and Brazil. Carrefour also has 230 convenience stores, mostly in Europe, a cash-and-carry operation that it appears to be gradually running down (twenty-three units by the end of 2008), and other diversified operations, particularly in Spain, where it owns travel, financial and insurance agencies.

Tesco has made remarkable progress in recent years, to become the outright leading retailer in the UK and, according to the IGD (formerly the Institute of Grocery Distribution), is poised to become the second largest global player after Walmart by 2012. The key to this, argues IGD, has been its international expansion, particularly into China and India, where it is expected to achieve a growth rate above 13 per cent over the next five years. Tesco Group sales reached £51.8 billion in 2008, an 11 per cent increase over 2007 and almost twice Carrefour's growth rate. This demonstrates the profitability to be gained from moving into emerging economies – sales from its Asian operations reached £6 billion in 2008, up 27 per cent on the preceding year, whereas sales increased by just 6.7 per cent in the UK to £38 billion. Tesco now accounts for £1 in every £3 spent by consumers on food in the UK, and has a range of store formats clearly designed to optimise the capture of consumer spending. This range extends from the giant Extra stores (60,000 square feet and above) through Superstores 20,000–50,000 square feet), the inner-city Metro stores (7000–15,000 square feet) and the small convenience stores known as Express (up to 3000 square feet), Tesco has also been at the forefront of innovations with regard to its loyalty-based "Clubcard"; making available personal financial services, including credit cards; internet-based shopping and home delivery; and mobile communications.

Metro Group is a German retailer with strong interests in wholesale cash-and-carry (655 stores in twenty-nine countries); Real stores, carrying food and non-food items, in 439 locations in five countries; Media Markt and Saturn, dealing in consumer electronics; and a department store, Galeria Kaufhof. Food consequently comprises only a modest part of its portfolio.

Kroger is a US company with a long history (125 years in business), and in its corporate publicity places some emphasis upon traditional values such as integrity and honesty. It operates 2481 stores in thirty-one US states under a variety of different names and formats, with supermarkets alone operating under sixteen different banners. It also has low-price warehouse stores, nearly 800 convenience stores, and 385 jewellery stores (described as "high margin business with good cash flow"). Kroger also operates forty food manufacturing facilities producing private label products on a three-tiered quality/price basis.

Sources: corporate websites, including the companies' published financial reports; Fishman (2006); IGD website (www.igd.com).

Food service

A striking feature of recent years has been the increasing proportion of consumer spending on food and drink consumed outside the home. The sale of food and drink for immediate consumption, either on the premises, in designated eating areas (e.g. those shared by different outlets in shopping malls), or to take away has consequently become a huge economic sector. People on the move are provided for at airports, stations and motorway services. Filling-station forecourt food sales are reputedly a more reliable source of profitability than the margins to be made from the sale of fuel, and testify to the significance of dashboard dining, grazing while on the move. Contract catering has become big business in company canteens, schools and hospitals, and in the hospitality industry. Developments in cook-chill and cook-freeze technologies, whereby specific ingredients to entire meals can be prepared in large volume by centralised operations and then transported to sites of consumption for reheating and serving, have become an important, if somewhat hidden, aspect of catering.

Food service is possibly one of the most dynamic and fast-moving sectors, responding quickly to changes in lifestyle and consumer demands. Europe accounts for the largest slice of this $730 billion industry, with almost one-third, while the US share amounts to 27 per cent and Asia-Pacific comprises 30 per cent. Of the total European food sector estimated at $210 billion in 2004, cafés, bars and restaurants accounted for 60 per cent, fast food 10 per cent and institutional provision 28 per cent. The latter category is interesting, for it is performed by businesses that are much less well known by the general public than fast food brands. Companies such as Sodexo, a provider of food service and facilities management in the USA, Canada and Mexico with turnover of US$20 billion in 2008; or the Compass Group, with headquarters in the UK, operations in fifty-five countries and revenues of £11 billion in 2008, both demonstrate the profitability of taking on contracts to provide the in-house delivery of food and drink for private businesses, hospitals or schools.

According to data from Defra, consumer expenditure on food service amounted to some £81 billion in 2008, slightly below the estimated £92 billion spent by UK households on food and drink through the retail sector (Defra 2009). Though both sectors have comparable levels of "gross value added", that is, the difference in the value of the goods and services produced and the cost of raw materials and other inputs, there are more than twice as many enterprises in the food service sector, employing almost 1.4 million people (compared with 1.1 million in retailing). While many of these might be dismissed as part-time "McJobs", it is clear that the food service industry catering to the provision of food and drink outside the home is not only a huge source of employment, but has performed a transforming role in shaping consumption patterns in the UK.

This is particularly evident in the fast food category. Here, possibly the best known brand is that of the McDonald's Corporation, which in 2008 generated revenues of $23.5 billion from its 31,000 outlets, more than three-quarters of which are run as franchise operations. Its main competitor, Yum! Brands – owners of KFC, Pizza Hut and Taco Bell, amongst other symbol chains – operates 36,000 outlets in 110 countries, generating revenues of more than $11 billion in 2008. Much like the supermarket retailers, there has been enormous expansion of these brands overseas, especially into the fastest growing economies such as China, where Yum! Brands alone has 3500 establishments. Clearly, this is an extraordinarily dynamic and innovative sector with an unparalleled level of global expansion. It is also one that would have long been considered a serious threat to building a more sustainable food system. Recently, however, some of the leading companies in the sector have engaged in public relations to demonstrate a commitment to sustainable practices, such as the national sourcing of some ingredients. It remains to be seen whether they will prove an ally in, or a major obstacle to, securing a more sustainable agri-food system.

8. Summary

This chapter has set out a structure for the global agri-food system and interrogated each of the major elements represented in Figure 2.1. It outlines how technological advances in agriculture have played a significant part in chaining farmers to a treadmill in search of higher productivity. This has been a feature not only across Europe, North America and other developed economies, particularly after 1945, but has become a process gathering pace across large swathes of the developing world since the 1970s. One of the key drivers of this treadmill is attributed to the way commercial interests effectively appropriated resources and technologies previously in the hands of farmers (e.g. seeds, soil fertility, traction) to become industrial inputs subject to innovation and scale enhancement. Meanwhile, downstream of the farm gate, food-manufacturing enterprises also enjoyed the benefits of accumulating value generated by the farm sector through contract production, substitutionist practices, and their creation of a global market for convenience foods. Though there remain, especially in some regions, large numbers of SMEs, the food processing and manufacturing sector has witnessed significant commercial consolidation. This has occurred both through horizontal expansion involving mergers, buy-outs and take-overs; and vertical integration involving the development of forward and backward links throughout the agro-food complex.

Now that the structure of the global agri-food system has been elaborated, it is possible to begin to examine its environmental dimensions. Chapter three explores the ecological basis of primary food production, examining the vital importance of ecosystem services and natural resources to farming. Chapter four then identifies four key challenges to primary production, issues that represent short- to medium-term threats to food security. Chapter five moves downstream and examines in a little more detail the environmental impacts associated with food manufacturing and contemporary retail practices.

Further reading

Weiss, T. (2007) *The Global Food Economy: The Battle for the Future of Farming.* London: Zed Books.
A concise text written by a Canadian geographer, it also sheds a particular light on the WTO and the multilateral regulation of agriculture.

Magdoff, F., Foster, J.B., Buttel F. (eds) (2000) *Hungry for Profit: The Agribusiness Threat to Farmers, Food and the Environment.* New York: Monthly Review Press.
A little dated now, but there are a number of excellent essays in this edited collection.

Galeano, E. (1973) *Open Veins of Latin America: Five Centuries of the Pillage of a Continent.* New York: Monthly Review Press.
This will not appeal to everyone: it is an utterly one-sided, totally gripping account of the brutality and injustice experienced by Latin America since 1492. This remains one of my inspirational books.

Mazoyer, M., Roudart, L.(2006) *A History of World Agriculture: From the Neolithic to the Current Crisis*. London: Earthscan.
A truly outstanding, encyclopaedic text that documents all kinds of agriculture around the world throughout time.

Tudge, C. (2003) *So Shall We Reap: What's Gone Wrong with the World's Food and How to Fix It*. London: Penguin.
Not only comprehensive in range, it explains in a thoroughly comprehensible way the biological basis of our food system. Tudge is an outstanding communicator of scientific principles to non-scientists.

Fernandez-Armesto, F. (2001) *Food: A History*. London: Macmillan.
A deeply scholarly account, but highly literary and engaging, F.F.A. is most evidently a gourmet as well as an outstanding historian.

And on the web

The Food Timeline: www.foodtimeline.org
One of the world's greatest websites. Not only a chronology, but a portal into a world of food history. Hours of browsing pleasure.

For critics of corporate food retailing
Walmart Watch: http://walmartwatch.com

And for observers of Tesco: www.tescopoly.org

③ The agro-ecology of primary food production

Contents

1. **Introduction**
2. **Ecosystem services**
3. **A typology of agricultural systems**
   Irrigated systems
   Rainfed systems
   Shifting cultivation
   Mixed farming systems
   Confined livestock and extensive grazing
   Aquaculture
4. **Resources for primary production**
   Soils
   Water
   Biodiversity
   Energy
5. **Summary**
   Further reading

Boxes

3.1 Millennium Ecosystem Assessment: main findings
3.2 Deforestation in Amazonia and soybean production
3.3 Total land requirements
3.4 Bolivia: potatoes and maize
3.5 Energy analysis
3.6 A comparison of energy inputs to land preparation

Figures

3.1 Framework to consider the range of primary food production options
3.2 A conceptual water accounting framework

Plates

3.1 Wet rice cultivation, West Sumatra, Indonesia
3.2 Pre-Hispanic terracing, Colca Valley, Peru
3.3 Agro-forestry: harvesting fruits, Lampung, Indonesia
3.4 Ducks for integrated pest management, West Sumatra, Indonesia
3.5 Aquaculture: cultivating fish in irrigation canals

Tables

3.1 Production statistics for five food and feed commodities, 2007
3.2 A global typology of primary food production systems
3.3 Energy inputs in US maize (corn) production

1. Introduction

The basis of the modern food system has involved major human interventions into natural ecosystems, and their fundamental redesign from complex and diverse habitats to **agro-ecosystems** often characterised by monoculture production. Such transformations have taken place in terrestrial (land-based) ecosystems to produce annual and perennial crops and livestock, as well as in coastal, estuarine and freshwater ecosystems in order to create highly productive aquaculture operations providing fin- and shellfish. This transformation of natural ecosystems got under way even before the development of agriculture around 12,000 years ago, when bands of hunters used fire to flush out their prey. Yet it is with the emergence of agriculture, when humans first planted, deliberately and consciously, a seed, tuber or cutting with the expectation that it would grow and produce a multiple of the original, that the composition of ecosystems began to be actively manipulated. This process appears to have developed independently at around the same time in several different parts of the world, including Mesopotamia (the Tigris–Euphrates Valleys of present-day Iraq); Eastern China; and Meso-America (Central America and Southern Mexico).

The displacement of natural vegetation by food crops was, for most of the past 10,000 years, an incremental process in keeping with the gradual rise in the human population. But from around 1700, with the development of the Industrial Revolution and European imperialism, this led to a faster rate of cropland expansion. With the growth of international trade, the grasslands of North and South America were ploughed and planted with wheat and maize, while conquered territory from New Zealand, through South-East Asia to Africa became sites for the production of new agricultural commodities (livestock, tropical beverages, fibres and rubber). Much of the past 300 years of ecosystem transformation has consequently been marked by a high degree of brute force and exploitation of natural resources and their local societies. Yet human ingenuity has not been entirely absent; indeed, in a technological sense, increasing scientific sophistication – for example, an ability to re-engineer the genetic composition of seeds, and precision farming using GPS – has come to symbolise human mastery over nature. Such developments have lulled many into complacency about our ability to exercise total dominion of the Earth, the oceans and the atmosphere, and thereby guarantee our future prosperity and food security. Yet such a perspective overlooks the degree to which we have altered the balance of nature, whether through the burning of fossil fuels and the destabilisation of the global climate system, the overfishing of marine stocks, or the unsustainable use of freshwater resources. Fortunately, there is a growing realisation of the degree to which our food production systems still rely upon the ecological services of nature as well as an appreciation that we need urgently to moderate our level of demands upon them.

This chapter focuses on the vital role performed by ecosystem services, and on the consequences arising from the neglect of environmental resources needed to support agriculture and other food provisioning systems on Earth. First, it is necessary to explain what is meant by ecosystem services and to establish the critical position of land as a "platform" for the provisioning of many of these services as well as all terrestrially derived food. The chapter then offers a typology of primary food production systems; although highly aggregated, the brief characterisations of each reveal their degree of dependence on ecosystem services. Finally, section 4 provides a more detailed appreciation of four key resources for primary production – soils, water, biodiversity and energy – and the discussion reveals a carelessness bordering on collective negligence in the sustainable management of these resources.

2. Ecosystem services

Between 2001 and 2005, the Millennium Ecosystem Assessment (MA) was carried out, involving the work of thousands of scientists and resulting in the publication of a number of reports. Its brief was "to assess the consequences of ecosystem change for human well-being and to establish the scientific basis for actions needed to enhance the conservation and sustainable use of ecosystems and their contributions to human well-being" (Millennium Ecosystem Assessment 2005a: v). Taking an ecosystem as "a dynamic complex of plant, animal, and micro-organism communities and the nonliving environment interacting as a functional unit", the MA was concerned with understanding how changes in ecosystem services influence human wellbeing. Ecosystem services are defined by the MA as benefits people obtain from ecosystems. These include: *provisioning services* such as food, water, timber and fibre; *regulating services* that affect climate, floods, disease, wastes and water quality; *cultural services* that provide recreational, aesthetic and spiritual benefits; and *supporting services* such as soil formation, photosynthesis and nutrient cycling (Box 3.1).

Box 3.1

Millennium Ecosystem Assessment: main findings

Over the past fifty years, humans have changed ecosystems more rapidly and extensively than in any comparable period in human history in order to meet growing demands for food, water, timber, fibre and fuel. This has resulted in a substantial and largely irreversible loss of diversity of life on Earth. Three major problems associated with the management of the world's ecosystems are as follows.

- Approximately 60 per cent (fifteen out of twenty-four) of the ecosystem services examined are being degraded or used unsustainably, including freshwater, capture fisheries, air and water purification, and the regulation of regional and local climate. Many ecosystem services have been degraded as a consequence of actions taken to increase the supply of other services, such as food. These trade-offs often shift the costs of degradation from one group of people to another, or defer costs to future generations.
- There is evidence that changes being made in ecosystems are increasing the likelihood of nonlinear changes (including accelerating, abrupt and potentially irreversible changes) that have important consequences for human wellbeing. Examples of such changes include disease emergence, abrupt alterations in water quality, the creation of "dead zones" in coastal waters, the collapse of fisheries, and shifts in regional climate.
- The degradation of ecosystem services is being borne disproportionately by the poor; is contributing to growing inequities and disparities across groups of people; and is sometimes the principal factor causing poverty and social conflict. The degradation of ecosystem services is already a significant barrier to achieving the Millennium Development Goals (MDGs) agreed by the international community in September 2000, and the harmful consequences of this degradation could grow significantly worse in the next fifty years. Rural poor people, a primary target of the MDGs, tend to be most directly reliant on ecosystem services and most vulnerable to changes in those services. More generally, any progress achieved in addressing the MDGs of poverty and hunger eradication, improved health, and environmental sustainability is unlikely to be sustained if most of the ecosystem services on which humanity relies continue to be degraded.

Source: adapted from Millennium Ecosystem Assessment
(2005a), 1–2.

Food production constitutes the largest user of ecosystems and their services:

● Approximately one-quarter of the Earth's terrestrial surface is now occupied by cultivated systems, which are defined as areas where at least 30 per cent of the landscape is in croplands or other food production systems (e.g. livestock or aquaculture).
● Cultivated systems have become the major global consumer of water, with major implications for river catchments and the needs of other water users
● The intensification of food production involving increased applications of agricultural chemicals has led to widespread deterioration in water quality, with increased contamination arising from nitrates, phosphates and other harmful pollutants.
● The expansion of cultivated systems has had a huge impact on most forms of wild biodiversity as a result of the destruction of native habitats and the use of pesticides to control invertebrates.

One of the most visible consequences of efforts to increase food provision has been the conversion of land use from natural habitat to that of cropland. In nine of fourteen terrestrial ecosystems identified by the MA, between 20 and 50 per cent of the areas once covered by primary vegetation have been cleared for crop cultivation. While this conversion may seem modest in light of the indispensable significance of food, this is to presume that these natural ecosystems are not performing vital services in their natural state. Developing a more holistic appreciation of the way agriculture is underpinned by ecosystem services is a key objective of this chapter.

From the outset it is important to recognise that the stock of land is limited, given that 71 per cent of the Earth's surface is covered by water. Of the approximately 135 million km^2 of land not covered by ice, around 11 per cent is under crops, 24 per cent comprises grasslands and 31 per cent forestry: the rest is either too inhospitable (mountains, deserts) or is already permanently marked by human occupation and used for housing, transport, industry, mining, reservoirs, waste disposal, recreation and so on (Clarke 1996).

Although the world witnessed a major expansion in the area under crop production from 1700, the MA notes that a greater amount of land was converted in the 30 years after 1950 than in the 150 years up to 1850. Alongside the faster rate of global population growth, the years after 1950 are also marked by rapid mechanisation and a greater technological capacity to undertake land-use conversion. Yet it is also important to note that, during the forty years to 2002, 71 per cent of increased food output in the South was achieved through increases in crop yield. Only in Sub-Saharan Africa did the share in the growth of crop production owe more to an expansion in cultivated area (66 per cent). Nevertheless, there remains considerable anxiety about the rate of deforestation in the humid tropics, particularly in Amazonia (Box 3.2).

Since 1950, the global population has doubled, yet food production more than kept pace, with cereals output increasing by 130 per cent. Such considerations do not prevent observers from speculating, however, about whether there is, or can be, sufficient land to support a projected increase of the global population to ten billion by around 2050. Indeed, a powerful current of neo-Malthusianism within the environmental literature has drawn attention to what it sees as an inevitable shortfall in the availability of arable and pasture land from which to meet the food needs of the growing global population. While it is vital to appreciate how the demand for food can compromise the sustainable functioning of ecosystem services, using the average Western diet as the norm by which to calculate total aggregate demand may not be the most appropriate way to proceed.

One way to address this question would be to create a database of all the present and potential arable and pasture land around the world, consider its productive capabilities, and compute how many people the Earth could support. An alternative perspective might begin from appreciating the impact of our present diets on land and explore how a moderation of our food consumption practices would help to reduce such pressures. Vaclav Smil has scrutinised such data and considers the total land area needed in relation to different dietary options. His observations are summarised in Box 3.3.

The conclusion one might draw from this is that, depending on one's perspective, we are reaching, have already reached or, indeed, have overshot the safe upper limits on the amount of land that might be converted from natural ecosystems to primary food production. The variable that is often overlooked in making such calculations is the composition of the human diet, especially the contribution of livestock products. This is a topic to which we return in chapters four and seven. In the meantime, it is important to be alert to possibilities for reconnecting our eating with the resource base, and to appreciate that dietary choices ought to be framed by an understanding of ecosystem

Box 3.2

Deforestation in Amazonia and soybean production

In its entirety, the Amazon Basin covers approximately 7 million km^2 – some 5 million km^2 in Brazil, the remainder in neighbouring states. At least 60 per cent of the world's remaining tropical rainforests, including an estimated 55,000 different plant species, are found in the Amazon. Forests in the Amazon Basin contain at least one-fifth the equivalent of all the carbon currently in the atmosphere, and are functioning as a globally significant carbon sink, mopping up some of the industrial emissions of carbon dioxide. However, carbon uptake by the intact forest is more than outweighed by carbon emissions from deforestation.

By 1998, the area of forest cleared in the Brazilian Amazon had reached some 549,000 km^2, about the size of France out of a total area as large as Western Europe. Given that the opening up of Amazonia only really began in the 1960s, in just a few decades, Brazil has cleared an area far greater than during the preceding five centuries of European colonisation. In 2003 alone, 23,750 km^2, an area the size of Belgium, was cleared. In 2004, remote satellite sensing picked up more than 35,000 separate fires in the Brazilian Amazon.

What drives this deforestation? Brazil's soybean production "has expanded at a pace rarely matched for a major crop in any country" (Brown 2005). In 1969, it produced 1 million tonnes (mt) of soybean and by 2004, 66 mt; the area under soy cultivation increased from less than 1 million ha to 24 million ha, with half this growth taking place since 1996. Soybean production spread from the southern states of Brazil into the *cerrado*, or savanna region, and is continuing to advance into the Amazon basin, displacing thousands of small, subsistence farmers, who then extend the agricultural frontier. Industrial soybean production is feeding a rapid escalation in factory farming, which has made Brazil one of the largest and fastest-growing producers and exporters of poultry and pig meat, and is propelling the growth of factory farms in other countries through the supply of soy-based animal feeds.

Brazil also has the world's largest commercial cattle herd, with 190 million head, and has overtaken the USA and Australia to become the largest beef exporter. Much of this beef is grazed on pasture sown in the ashes of the Amazon forest.

Sources: Brown (2005); Bunyard (2007); Weiss (2007).

Box 3.3

Total land requirements

In a world of 9–10 billion people, total land requirements will depend on the composition of their average diets and on the intensity of cultivation. For a largely vegetarian diet produced by high-intensity cropping, no more than 700–800 m² per capita would be required. The existing Western diet, with its high share of meat and dairy products, claims up to 4000 m² per person, though this includes a significant share of food that is wasted.

Considering that an average daily supply of 2500 kcal per capita would be quite adequate for affluent nations, and if the 30 per cent share of animal products were composed largely of dairy products, poultry and pork, then even a moderately intensive single cropping would not need more than about 1500 m² per capita to provide it. In contrast, a high share of beef in the diet would push this up to 3000 m² per capita. Consequently, the land requirements needed to feed 10 billion people would range from 800 million to 3000 million ha, and the most reasonable range would be between 1100 and 1500 million ha. This would call for no more farmland than we already cultivate, even without further intensification of cropping. The conclusion is that, in global terms, the availability of farmland is not a limiting factor in the quest for decent nutrition during the next two generations.

Source: adapted from Smil (2000), 38.

---

services. This need not be a constraint to the potential variety of foods that we can eat: on the contrary, there are strong reasons to suggest a widening of the range of foods in the interests of biodiversity conservation.

3. A typology of agricultural systems

Despite phenomenal growth in the oil crops sector over the past forty years (as illustrated by the Brazilian example described in Box 3.2), which is a consequence of changes in consumption – most especially of meat, processed foods and snacks – cereals remain the most important category of food crops. Cereals provide around half of calories consumed directly by humans and occupy nearly 60 per cent of the world's harvested crop area, while accounting for a greater share of fertiliser, water and other agri-inputs. The cereals sector is dominated by the big three – rice, maize and wheat – with sorghum, millet and barley coming some way behind in terms of the volume of global output. Of course, the overwhelming thrust of agricultural research effort has been directed to the first three crops, although sorghum and millet (and t'ef, and quinoa) are all regionally significant grains well adapted to their local agro-ecological circumstances. Production data for the three main cereals, soybean and potato, disaggregated by world region, are shown in Table 3.1.

Table 3.1 *Production statistics for five food and feed commodities, 2007*

Commodity	Area harvested (million ha)	Production (million tonnes)
Wheat		
World	214.2	606.0
Africa	9.5	18.9
Americas	38.8	103.0
Asia	97.4	281.1
Europe	56.1	189.7
Oceania	12.4	13.4
Maize		
World	158.0	791.8
Africa	28.0	47.7
Americas	65.7	454.5
Asia	50.4	217.0
Europe	13.9	72.1
Oceania	0.07	0.4
Rice		
World	155.8	659.6
Africa	9.1	21.3
Americas	6.7	33.6
Asia	139.4	600.9
Europe	0.6	3.6
Oceania	0.03	0.18
Soybean		
World	90.2	220.5
Africa	1.2	1.3
Americas	67.6	189.4
Asia	19.5	27.2
Europe	1.9	2.6
Oceania	0.01	0.03
Potato		
World	18.6	309.3
Africa	1.6	18.0
Americas	1.6	41.6
Asia	8.0	116.5
Europe	7.2	131.6
Oceania	0.045	1.7

Source: compiled from Food and Agriculture Organization (FAO) data, FAOSTAT, http://faostat.fao.org.

Maize has become the most important cereal by total volume of production (almost 800 million tones in 2007), though its recent expansion owes less to demand for it as food or livestock feed than as biofuel (an issue discussed in chapter four). Although it is still accorded sacred value in parts of rural Mexico, maize has become an utterly utilitarian material with multiple uses throughout the food industry, especially as feedstock for its conversion to high-fructose corn syrup, an industrial sweetener, and as animal feed. Nevertheless, it remains a staple food not only in much of Mexico and Central America, but to an extent further south, and also throughout Southern and East Africa and the Philippines.

Wheat is the most important crop in terms of area planted, and increasingly prevalent due to dietary change. The "westernisation" of consumption in the South, in which bread (burger buns, "sub" sandwiches), snack foods and pasta, all containing wheat, are becoming more ubiquitous, is a feature of many societies once far removed from wheat-cultivating cultures. Concerted scientific plant breeding has enabled wheat not only to continue to push forward the northerly limits of its cultivation across the Russian steppes and Canada, but also to make it more drought-tolerant and capable of performing well in northern Mexico and the Punjab region of India and Pakistan.

Finally, rice is the third major cereal and, since the dawn of agriculture, has proven the single most important staple food across East, South-East and South Asia. Indeed, Asia accounts for 91 per cent of global rice production. Although we typically associate rice with its cultivation in paddy fields (the most productive farming system, capable of producing up to three crops per year), upland varieties are widely grown provided there is sufficient rainfall and warmth.

Livestock and their products constitute a significant and growing share of the total gross value of agricultural output, comprising around half in the North and a third in the South, but the latter proportion is rising rapidly. Because of rising demand for livestock products, industrial systems of production, known as factory farms or confined animal feeding operations (CAFOs), are increasingly the norm in pork, poultry and egg production, with increasing amounts of feedlot beef and dairy. Well over 20 per cent of global food crop production is now destined for animal feeds, with a significant proportion comprising high-protein processed soybean from countries such as Brazil. The degree of development of the global livestock feed (and oil crops) industry is indicated by the fact that 86 per cent of global soybean production now occurs in the Americas (with 94 per cent of this in the three largest producers – USA, Brazil and Argentina) compared with 12 per cent in Asia. Soybean cultivation originated in China and, as we know, far eastern cuisine makes extensive use of soy products. Yet, despite China being the largest global producer of rice and wheat, and second largest of maize, its soybean production amounts to just 6 per cent of global output and less than one-quarter of Brazil's production (FAO data, http://faostat.fao.org).

The diversity of agricultural systems, comprising different combinations of crop and/or livestock production, has long engaged geographers in devising classification schemes through which to order and map these systems. Many have been based on the physical properties of land or its capabilities, creating regional patterns reflecting a preponderance of one type of production. Other disciplinary approaches construct different schemes based upon socio-economic parameters they would consider to be of greater importance: tenure and extent of landholding; degree of intensification and product specialisation; level of technological innovation, including energy inputs; income (wealth); the nature and form of social organisation, and so on. Consequently, there are a potentially enormous variety of ways to classify food production systems although, ideally, they should acknowledge ecological parameters as well as social, economic and cultural criteria. A framework through which to consider the potential diversity of primary food production systems (focused largely on arable farming) is shown in Figure 3.1.

In order to avoid detailed regional descriptions of many different farming systems, yet wishing to illustrate the links between food production and ecosystem services, this chapter draws upon the MA, which offers a useful typology of agricultural systems. It is a highly aggregated classification, constructed with regard to broad agro-ecological parameters. Horizontally, the agro-ecological characterisation is simplified and, in the first instance, broken into two broad categories: tropical and subtropical, comprising 62 per cent of the total cultivated area; and temperate, comprising around 38 per cent, with further subdivision as shown in Table 3.2. The criteria that determine these categories include day length, solar radiation and temperature, and moisture regimes (taking into account rainfall and evapotranspiration).

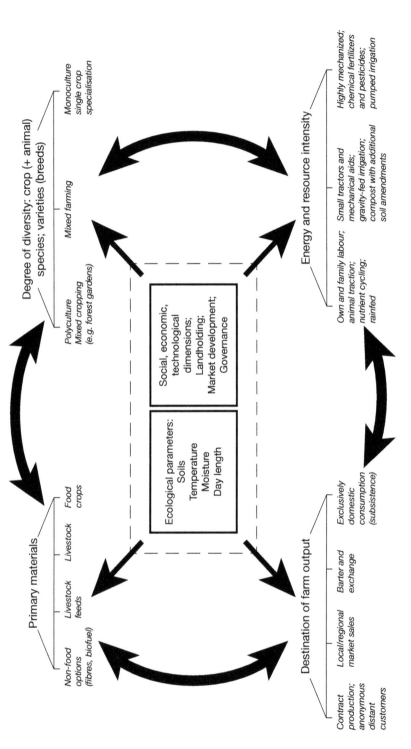

Figure 3.1 A framework for evaluating *primary food production options*

This typology is represented as a matrix in Table 3.2. Each farming system is then briefly described in turn.

Irrigated systems

Although irrigated agriculture constitutes around 18 per cent of the total cultivated area, it provides about 40 per cent of total crop production, due to double or even triple cropping of land where water availability and temperature permit such intensification. The central feature is the management and delivery of water to cultivated fields, sometimes involving the installation of massive engineering structures (dams) and hundreds if not thousands of kilometres of canals. Elsewhere, irrigation may be drawn from groundwater using electric or diesel pumps, or animal or even human labour in wells, hand-pumps or Archimedes screws. Some of the finest examples of irrigated agriculture are the lowland paddy rice-farming systems of South-East Asia, where small-scale irrigation schemes managed by local communities have a generally distinguished record of maintaining a highly productive and sustainable system. Here, the creation of terraces on even gently sloping land optimises water use, with gravity and human ingenuity working in harmony to deliver crop moisture to a succession of fields on a gradual fall across the contours (see the series of images illustrating aspects of wet rice cultivation in West Sumatra in Plate section 3.1). However, poorly maintained and badly managed irrigation schemes elsewhere have resulted in serious problems of soil degradation, while those systems dependent on the extraction of groundwater can result in the depletion of that resource if rates of use exceed the rate of replenishment. The serious challenge of managing water resources worldwide is addressed below, and in greater detail in chapter four.

Rainfed systems

According to the MA, rainfed agriculture accounts for about 82 per cent of the total cultivated land area and exists in all parts of the world. Generally, and providing temperatures permit, the reliability, quantity and distribution of rainfall will determine the kinds of farming systems that develop within a region. Areas of high agricultural potential, that is, with adequate rainfall and fertile soils, generally utilise high external-input systems to produce annual crops such as the intensive cereal farming of Western Europe or North America. Elsewhere, perennial crops and/or livestock may form part of a mixed farming system (see

Table 3.2 A global typology of primary food production systems

Farming system	Tropical and subtropical (62 per cent of global cultivated area)			Temperate (38 per cent of global cultivated area)	
	Warm humid/subhumid	Warm semiarid/arid	Cool/cold highland montane	Humid/subhumid	Semiarid/arid
Irrigated	Rice (East, South-East Asia); rice–wheat (Pakistan, India)	Rice (Peru); wheat, rice, cotton (Egypt)	Maize, beans (Andes)		
Rainfed, high external input (crops, livestock)	Rice–pulses; maize, sugarcane		Tea, coffee plantations (East Africa, Sri Lanka)	Maize–soybean (USA, Brazil, Argentina)	
Rainfed, low external input (crops, livestock)	Tropical staple crops (yams, cassava, banana in Sub-Saharan Africa)	Mixed crop livestock (Sahel, Australia)	Tubers, quinoa (Andes)		Wheat–fallow systems (Central Asia, Australia)
Shifting cultivation	Important in Amazon basin, Central America, South-East Asia				
Extensive grazing systems		Important in East, Southern Africa	Alpaca, llama (Andes)	Uplands of Atlantic Europe (sheep, goats)	
Industrial confined livestock systems	Exist independently of agro-climatic circumstances				
Aquaculture	Freshwater: carp, tilapia (China, South-East Asia); coastal margins: shrimp (Thailand)			Salmon farming (North-West Europe, north-west coast, USA and Canada)	

Source: adapted from Millennium Ecosystem Assessment (2005b): ch. 26.

below). As rainfall amounts decline and/or become increasingly seasonally concentrated, agricultural potential diminishes and leads farmers to invest less in external inputs. Annual crops may be marked by their ability to withstand drier conditions (drought tolerance) and include sorghum, millet and groundnut. Indeed, perennial (tree) crops may represent a more reliable cropping system, or, as conditions become increasingly arid, livestock that can browse on native vegetation. Clearly, rainfed agricultural systems encompass a wide spectrum of

Plate section 3.1 *Aspects of wet rice cultivation, West Sumatra, Indonesia*

continued overleaf . . .

Plate section 3.1 *continued*

faming possibilities and, together with other agro-ecological parameters, principally temperature, determine the kinds – and quantities – of foods that can be produced. Rainfed systems in hill and mountain areas may give rise to considerable investment in land engineering works such as terracing, particularly where there are high population densities (mouths to feed, but also the labour power to perform such work) and the possibilities of capturing and channelling seasonal irrigation (e.g. from snow or glacier meltwater). Plate section 3.2 shows the extent of pre-Hispanic (Inca-era) terracing in the Colca Valley, Peru, when the population was higher than at present.

Shifting cultivation

Shifting cultivation, or "slash-and-burn" agriculture, is probably the oldest, most widely practised and amongst the most controversial types of farming. It is most commonly found in and around forested environments throughout the tropics, and involves clearing plots of land ("swiddens") by felling trees and burning the vegetation. Depending on the prevailing food culture, crops might include rice, maize, cassava and yam/sweet potato. Under low population densities, this system works relatively well, with plots cleared and planted for two or, at most, three years before being allowed to revert back to natural vegetation. After twenty-five or thirty years in tropical moist forest environments, secondary regrowth is almost indistinguishable from primary forest, and the cycle can begin again. However, as population densities increase, the availability of primary or long-fallowed secondary forest is limited, and swiddens are cleared before regeneration is complete. Such environments are usually characterised by poor soils with most of the fertility locked up in growing vegetation and released in the process of burning. Once the nutrients left in the ashes of the forest have been taken up by a growing crop, or more significantly, washed away by heavy tropical rains, there is little left to support secondary cultivation.

Opponents of this farming system include governments, which associate such practices with unassimilated ethnic groups (a situation that prevails in northern Thailand, Myanmar and south-west China), as well as those more widely concerned with loss of biodiversity and the potential contribution to climate change (see below). Nevertheless, permanent cultivation on such poor tropical soils is extremely difficult, unless there is some attempt to recreate the architecture and diversity of the original forest, and here a form of cultivation known as agro-forestry works best, though it requires effort to develop the most appropriate – culturally and

Plate section 3.2 *Pre-Hispanic terracing, Colca Valley, Peru*

biologically – combinations of plant species. Plate section 3.3 shows a method for harvesting produce in an agroforestry system comprising coffee, pepper, papaya and other perennials.

Mixed farming systems

This represents the most significant number of farms, where there is a mix between annual crops and livestock. Mixed systems exist in both irrigated and rainfed contexts, across the full spectrum of agro-ecological environments, and offer real synergies, both economically and environmentally, for smallholder farmers. For example, larger animals (cattle, buffalo, horses) in many different societies have been indispensable in performing field tasks (e.g. ploughing), while these together with other species (mules, donkeys, llamas, camels) have proven invaluable for carrying or pulling agricultural produce to market. Their manure, together with that of other grazing animals, may be carefully husbanded as a source of soil fertility or, mixed with straw and dried, to provide fuel with which to cook food for household consumption. Lactating animals provide a nutritious source of food for young and old (milk), which can also be preserved (cheese). In return for all these valuable services, animals may graze on crop residues (field stubble), utilise common pasture, or browse around field boundaries and roadside verges. Where dairying becomes more important, farmers may practise "cut-and-carry" feeding systems with tethered or confined animals.

Non-ruminant animals (those not able to digest high cellulosic grasses) may rely more upon scavenging, supplemented by kitchen wastes. Pigs, except in those countries where pork is regarded as culturally offensive, are widely kept for their relative efficiency in converting scraps to meat and fat. Poultry, too, offers the prospect of eggs and meat for little outlay or effort. In wet rice farming systems, ducks also make an important contribution to integrated pest management, dibbling through paddy fields after the harvest of one crop for the larvae of insects that would otherwise attack the subsequent rice crop (see Plate section 3.4). It is little wonder that mixed crop and livestock systems have proven so adaptable and beneficial in so many different parts of the world.

Confined livestock systems and extensive grazing

Although we might extol the virtues of mixed farming, in practice the demand for livestock products has long outstripped the capacity of such

Plate section 3.3 *Agro-forestry: harvesting fruits, Lampung, Indonesia*

systems to supply them. Consequently, industrial systems of meat, dairy and egg production have emerged designed to maximise productivity in the shortest possible time frame. Such systems can be considered industrial insofar as they supply all inputs, comprising concentrated feed rations and prophylactic antibiotics designed to control infection amongst animals confined at high densities, with the purpose of optimising weight gain and yield increase. Such systems now account for most of the world's poultry (egg and broiler chicken) and pig meat production, and an increasing proportion of dairy in the developed countries is derived from stall-fed cattle. Intensive feedlot operations have been a feature of beef production in the United States, and this model is being disseminated more widely as the desire for cheap meat takes hold.

As the MA report observes, while "there are good economic arguments for the concentration of large numbers of animals associated with many confined systems, there can be significant impacts on surrounding ecosystems" (Millennium Ecosystem Assessment 2005b: 752). The disposal of large amounts of manure and slaughter by-products has become extremely problematic in some regions, with soil capacity quickly saturated with nitrogen and phosphorus, and further applications resulting in nutrient contamination of water courses, an issue to which we return at the end of chapter four.

While confined livestock-feeding systems appear to operate independently of the prevailing agro-ecological conditions, extensive grazing systems, in contrast, have arisen largely as an effective way of utilising the more challenging environmental circumstances of drylands and temperate uplands. According to the MA, arid rangelands support around 50 per cent of the world's livestock and provide livelihoods to many millions of people who engage in mobile grazing practices. Although traditionally well suited to the resource base, efforts by government to sedentarise pastoralists have led to more localised grazing pressure around waterholes and planned villages.

Aquaculture

Terrestrial and aquatic food production systems differ in fundamental ways, as Brander (2007) has noted. Most terrestrial food production derives from farming systems, where selected crops are grown under controlled conditions, usually with additional fertiliser and the removal of predators and pests, and similar conditions prevail for domesticated livestock. Both crops and animals have been subject to genetic selection

Plate section 3.4 *Ducks as part of integrated pest management, West Sumatra*

and enhancement. Aquatic food production systems, on the other hand, have until recently been dominated by the capture of wild stocks. When the means for harvesting wild populations was relatively rudimentary and, most importantly, not far from shore, stocks could replenish themselves outside the range of fishing operations. However, with the arrival of industrial-scale fishing across the world's oceans, wild fish stocks have gone into serious decline. Although this has been masked to some extent by fleets switching to other species ("fishing down the chain"), it is now evident that at least 70 per cent of the world's fish stocks are estimated to be fully exploited, overexploited or recovering from depletion (Brander 2007, Pauly *et al.* 2002). It is for this reason that aquaculture has grown in scale, extent and commercial importance.

Aquacultural systems exist in coastal, estuarine and saline wetlands, as well as inland using freshwater resources. Many different production systems have been developed through which to rear and farm finfish, crustaceans and molluscs, and large-scale commercial examples of these are emerging in all parts of the world. Aquaculture now accounts for about one-third of the world's total provision of fish. Farmed seafood has been growing rapidly, largely as a response to increasing consumer demand for a healthier protein source, which has occurred at the same time as the collapse in stocks of the most popular wild fish species. The growth in salmon farms along the Pacific coast from Chile to Canada and Alaska, and around the Northern European Atlantic fringe (Norway, Scotland, Ireland), might appear an obvious commercial response to consumer demand. However, the feed requirements for this predator have led to enormous pressure on stocks of smaller wild species such as anchovy and whiting, which are harvested in order to generate high-protein fish-meal feed. Large-scale salmon farming has also come under increasing criticism for the ecological costs associated with pollution caused by unused feed and manure, as well as the spread of disease and contamination of the wild gene pool arising from the escape of genetically modified (GM) salmon.

Another aquacultural system that has caused considerable concern is the production in a number of Asian countries, primarily for export, of high-value shrimp. Creating suitable production facilities for shrimp farming has involved the removal of large areas of mangrove forests, these being a habitat comprising trees and shrubs adapted to living permanently or periodically in saline water. Mangroves perform truly vital ecological services, buffering and protecting the coastal margins from storms and typhoons, reducing exposure to flooding and erosion, and trapping sediment and limiting erosion.

Freshwater aquaculture is most widely practised in Asia and here two species, carp and tilapia, dominate. As herbivores, these fish can thrive on more locally available feeds and can be raised in a variety of different systems: in ponds, or within cages or enclosures in lakes, streams or irrigation canals. Such practices can play an important role in rural areas, providing a significant protein contribution to local diets and even a small income stream from sales of surplus fish. However, aquaculture that utilises common-property water resources, while illustrating the possibilities of multifunctional use, does require the cooperation of other users. Evidently, caged fish will not thrive where streams carry pollutants such as detergents from the washing of clothes or agricultural contaminants. Plate section 3.5 offers illustrations from West Sumatra, Indonesia.

This section has only briefly outlined some of the main types of farming systems worldwide. It is recognised that this characterisation represents a high level of abstraction; but given that the purpose of this book is to examine more broadly the links between food and environment, there is simply not space to engage in a more detailed analysis. The following section turns to examining the role and "state" of key resources for primary production: soils, water, biological diversity and energy.

Plate section 3.5 *Aquaculture: cultivating fish in irrigation canals, West Sumatra*

Plate section 3.5 *continued*

Plate section 3.5 *continued*

4. Resources for primary production

Soils

Soils form over *time* by the action of *climate* and *living organisms* on the underlying geological *parent material* (bedrock). The decomposition of plant materials provides organic matter. Soils vary enormously around the world according to the differences in these four soil-forming factors. They also vary according to their position in the landscape (e.g. on hillsides and valley bottoms), which is the factor of *topography*. Soil fertility is generally a term used to describe how well a soil can support growing plants or crops, and involves physical, chemical and biological properties.

Soil is one of the really remarkable living communities on Earth, with healthy soil lying at the heart of a nourishing food system. The notion of the soil food web helps to capture the complex ecology that ultimately starts with decaying plant matter (leaf litter, crop residues). The first level in the feeding chain comprises bacteria and fungi, which glue soil particles together, store nutrients and release them in a form for plants to feed upon. Above these are protozoa, arthropods (insects) and earthworms, then higher predators such as birds and small mammals. This complex ecology is not just a medium for growing food, but sequesters carbon, stores and filters water, and provides other valuable environmental services.

Yet for many farmers today, soil is simply a utilitarian medium in which to grow profitable crops and a material that can be manipulated and re-engineered to reduce costs and increase yield. Ploughing of soil has been practised for thousands of years, with livestock (oxen, buffalo, horses and camels) used before the rise of oil-powered tractors. Since the digging stick, tilling the land has been an indispensable feature of farming, opening the furrow to receive seed, and other operations performed to eliminate weeds and pests. Yet the growing of annual row crops that, when removed, leave the soil exposed to rain and wind; the cultivation of slopes without appropriate conservation methods; and the use of heavy machinery resulting in compaction of the soil have all contributed to high rates of erosion. Moreover, excessive reliance upon chemical fertilisers to provide a quick nutrient fix has led us to undervalue the importance of soil organic matter, while applications of pesticides have steadily reduced the complexity of the soil food web.

It has been suggested that at least 75 billion tonnes of soil is lost worldwide each year and that 80 per cent of the world's agricultural land suffers moderate to severe erosion. During the past forty years, about 30 per cent of the world's arable land has become unproductive, and much has been abandoned for agricultural use. Indeed, an estimated 2 billion ha of arable land has been abandoned by humans since farming began – a frightening amount when one considers the area of arable land actually under cultivation is around 1.5 billion ha (Pimentel and Pimentel 2008). Each year, an estimated 10 million ha of cropland worldwide are abandoned due to lack of productivity caused by soil erosion, with the highest losses in Asia, Africa and South America. In these regions, small farms are often located on marginal, steeply sloping land where farmers cultivate annual row crops. Yet even in the USA, with its experience of the Dust Bowl years of the 1920s and 1930s when winds carried away the exposed top soil of the Great Plains, high levels of erosion continue in states such as Iowa, where land is left uncovered during winter months leaving it vulnerable to the agencies of wind and water.

While soil conservation policy in the USA has largely been driven by a need to achieve tolerable soil loss, or the T factor – the maximum annual soil loss that can occur on a particular soil while sustaining long term agricultural productivity – there are now calls to "go beyond T and manage for C" (that is, carbon). In other words, it is now time to recognise that a preoccupation with erosion control measures (terracing, contour strips, cover crops) is not enough, and what is needed is greater, more holistic appreciation for soil *quality*. Recognising the critical ecological functions of healthy soil – especially its water, nutrient and carbon-holding capacity, its potential to reduce atmospheric CO_2, and its capability to produce nutritious food – is a vital first step. Thereafter, efforts to enhance soil organic matter will involve substantial improvements in land management practices, particularly the frequency of tillage operations and the applications of agri-chemicals (Warshall 2002).

The harvest of food crops from the field, often involving removal of the entire plant biomass, is effectively a means of extracting nutrients from the soil, and these must be replenished if fertility is to be maintained. The importance of synthetic fertilisers to global food production was discussed in chapter two, in which the role of nitrogen in particular was highlighted. The three macro-nutrients (nitrogen, phosphorus and potassium) may need to be applied to fields in tens (P and K) to hundreds (N) of kilograms per hectare every year. Yet it has been well documented that, at least until recently, only 30–50 per cent of nitrogen fertiliser and around 45 per cent of phosphorus applied to fields is actually taken up by crops. This means that up to half the macro-nutrients added to the soil are lost, finding their way through a number of pathways into the wider environment.

Micro-nutrients, including such minerals as calcium, zinc, manganese and boron, that need to be available in the soil for optimum plant growth may require applications rarely exceeding a few hundred grams per hectare, and this not every year (Smil 2000). Despite the small quantities involved, their widespread deficiency in soils used for growing food crops has been attributed as a cause of nutritional difficulty, degenerative diseases and ill health (Harvey 2006). In pre-industrial societies, farmers added organic matter – livestock manure, seaweed, fish heads – and practised a rotation of crops including clover, beans or alfalfa, which were known to help fix nitrogen through their root nodules that harbour rhizobia. With the intensification of soil use with industrial farming, significant additional sources of nutrients were needed, most especially nitrogen, and a preoccupation with ensuring the availability of this key element to plant growth caused a narrower and more functional approach to soil.

The contrast in attitudes to soil – between the utilitarian view that believes it to be simply a matter of supplying the macro-nutrients NPK, and a more holistic approach that views plant nutrition as more than the supply of soluble nutrients – is not a recent controversy. Fully seventy years ago, Sir Albert Howard set out the principles for the restoration and maintenance of soil fertility in his book *An Agricultural Testament* (Howard 1943). His view was that soils represented a stock of natural capital for each nation, and that there was a responsibility to utilise and to safeguard this asset. For Howard, the connection between a fertile soil and healthy crops, healthy animals and healthy human beings was indisputable; and with soils slowly being poisoned by the addition of artificial fertilisers, the growth of disease in plants, animals and humans was evidence of this. This was a view shared over fifty years ago by Lady Eve Balfour, the founder of the Soil Association, the organisation that represents and certifies organic farmers and growers in the UK. It is also a position increasingly being taken up by those who subscribe to the need to re-establish a more sustainable agri-food system and who recognise that we must reprioritise the value accorded to the health and integrity of soils.

Water

Water is the infrastructure for food. Water enables plants to grow and nurtures all living animals; it is used throughout food chains to prepare, cook and accompany food. Clean water is essential for drinking, for domestic use and for bodily functions. Without water, diet is dramatically curtailed: no water, no life.

Generally, the moisture requirements of crops are expressed in relation to potential evapotranspiration (PET), which is the quantity of water evaporated from the soil and plant surfaces when they are permanently moist. If the ratio of rainfall to PET is 1, a balance exists between the two. If it is less than 1 (e.g. 0.7 or 70 per cent of water requirements met), the degree to which this shortfall in moisture will affect plant growth will depend on the time of year and stage in the crop cycle. Clearly, if annual precipitation is not sufficient to meet the PET of those crops being grown in a region, irrigation is needed to increase the availability of moisture to them. This is achieved by withdrawing water from surface (streams, rivers, lakes) or groundwater resources.

In addition to our need for a litre or more of drinking water per day, it is estimated that we *eat* between 1600 and 5000 litres of water per day depending on our diet. One hectare of wheat will transpire more than 5–7 million litres of water each growing season, while an additional 2 million litres may be lost by evaporation from the soil (Pimentel and Pimentel 2008). Consequently, 1 kg of wheat takes between 500 and 4000 litres of water, while 1 kg of beef takes at least 10,000 litres. In total, it has been estimated that somewhere around 5200 km^3 of water is embedded in global food production (Hoekstra and Chapagain 2007).

Yet only 3 per cent of the water on Earth is in the form of freshwater suitable for drinking and for agricultural use, and much of this is locked up in glaciers and ice caps. An estimated 110×10^{12} cubic metres of rain falls on the Earth's surface, but only a little more than 10 per cent of this is usable. Freshwater is consequently precious – as each of the estimated 1.4 billion people on Earth currently lacking access to adequate safe water understand. Yet there is enormous variability in access to freshwater between regions and countries, and this has a significant influence on the potential for agriculture, for economic development more generally, and consequently for human wellbeing.

For example, in Africa 206 m^3 of water per person per year is withdrawn for agriculture, and this represents around 85 per cent of total water withdrawals on that continent. In North America, by contrast, the 1029 m^3 per person withdrawn for agriculture each year represents just 47 per cent of the total there (Millennium Ecosystem Assessment 2005b). Compared with much of the high-income North, many of the poorest countries tend to have scarcer water supplies and a greater need to use them in agricultural production, especially given the disproportionately greater role played by this sector in their economies. However, even within Europe there are significant variations between regions in levels of water extraction and the purposes to which it is put. If total abstraction of freshwater across Europe is around 500 m^3 per capita per year, in Eastern Europe more than half is used for cooling in electricity generation plants, while in Southern Europe more than 60 per cent is used for agriculture, rising to 80 per cent in some regions. Moreover, while nearly all water for power generation is returned from whence it was extracted, if a little warmer, barely one-third of water extracted for agriculture finds its way back into the environment, and this is more than likely contaminated with nitrates or other agri-chemicals.

The cultivation of food crops is consequently the single largest influence on the quantity and quality of freshwater resources, and this is especially apparent with regard to irrigation. As noted earlier, irrigated agriculture occupies just 18 per cent of global farmland but produces 40 per cent of crop output. In Spain, just 14 per cent of farmland is irrigated but this yields 60 per cent of the total value of agricultural products. It is little wonder, then, that there has been considerable pressure to extend the area under irrigation, which grew nearly fivefold during the past 100 years, with new schemes involving ever-larger feats of water engineering. Yet if the twentieth century witnessed the apogee of the "hydraulic mission", in which the engineering paradigm of dominating nature blossomed fully, leaving a legacy of 50,000 large dams, recent decades have begun to pay attention to the enormous social and environmental costs of such interventions (Molle *et al.* 2008). The discussion of water footprints in chapter four provides further examples of some of the more problematic aspects of this large-scale water engineering approach.

Figure 3.2 provides a conceptual water-accounting framework. This simple model illustrates how water enters and leaves a hydrological system, part of which is used to support agriculture. Note that decisions about what to produce in an area will depend to a significant degree upon the balance between the different water inputs and the transpiration

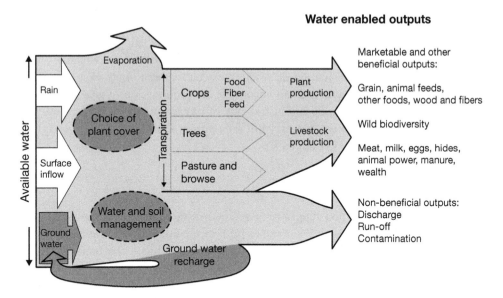

Figure 3.2 *A conceptual water-accounting framework*
Source: Millennium Ecosystem Assessment 2005c: 197.

requirements of the different crops and animals. Where rainfall is insufficient and surface flow limited, a greater amount must be drawn from groundwater resources. We should remember, too, that the water demands of food production compete with the needs of other users. Although water that is surplus to the requirements of growing crops may be returned to surface flows or recharge groundwater reservoirs, under intensive production there is considerable risk of such water being contaminated by animal wastes or agri-chemical runoff and rendered unsuitable for other water users.

As noted earlier, if less than half of all applied fertilisers are taken up by crops, this means that the remainder of the macro-nutrients added to the soil are lost, finding their way through a number of pathways into the wider environment. Nitrogen is most commonly leached in a soluble form as nitrate (NO_3) and in gaseous form as ammonia (NH_3), and nitrous oxide (N_2O), a greenhouse gas that contributes to the process of global warming. Nitrate accumulation in groundwater is a major source of concern in areas of intensive agriculture such as Western Europe and the American mid-west, where recommended limits on its concentration are frequently breached. High nitrate levels in drinking water pose risks to human health, including hampering oxygen circulation through the blood stream, which is especially dangerous in young children, a condition known as "blue baby syndrome". Nitrates are also associated with increased risk of stomach and other cancers. High levels of dissolved nitrates and phosphorus cause eutrophication of streams, rivers, estuaries and lakes, where these nutrients promote excessive plant growth and decay and can encourage the proliferation of algae. Subsequent decomposition and lack of oxygen in the water creates a toxic environment that may kill fish, poison shellfish and become hazardous to other animals.

Biodiversity

The world's food supply rests upon an extraordinarily narrow genetic base. According to Diamond (1998), of the 200,000 wild plant species only a few thousand are eaten by humans, and just a few hundred of these have been domesticated. Of these several hundred, most provide minor supplements to our diet while three-quarters of the world's food comes from only seven crop species – wheat, rice, maize, potatoes, barley, cassava and sorghum. Nearly half of the world's calorie and protein intake eaten as food – not converted as feed to livestock products – comes from only the first three cereal crops. Adding one pulse (soybean), one tuber (sweet potato), two sugar sources (cane and beet) and one fruit (banana) to the list of seven would account for over 80 per cent of total crop tonnage (Hawken *et al.* 1999). According to Pimentel and Pimentel (2008), 90 per cent of the world's food supply comes from only fifteen species of crop plant and eight species of livestock, which is an especially narrow base considering their estimate of 10 million species of plants and animals in the world today.

However, not only is the superstructure of the agri-food system built upon rather limited genetic foundations, these are narrowing further as a result of agricultural modernisation and intensification. As discussed in chapter two, prior to the Second World War, farming systems in countries of the North were overwhelmingly mixed, multifunctional operations combining crops and livestock, with both marked by some species and breed diversity. Throughout much of the rest of the world, and especially in the tropics – at least outside the colonial plantations – agriculture was characterised, above all, by diversity. Here, rural communities from the Andes to Asia via Africa maintained the legacy of locally adapted landraces purposively selected and bred on the grounds of at least two major functions. First, maintaining many varieties was a strategy to minimise risk and make the best use of uncertainty (poor or late rains) and hazards (outbreaks of pest or disease).

Consequently, practices such as polycultures, intercropping, staggered planting times and so on had the advantage of reducing total crop losses, spreading labour demands, diversifying food sources and ensuring some degree of food security over the year. Second, different varieties were cultivated for different culinary and cultural purposes: maize, for example, can be prepared to produce many different kinds of food and beverage (see Box 3.4).

Box 3.4

Bolivia: potatoes and maize

In the highlands of Bolivia, potatoes and maize are the key food staples of rural communities, though the contribution of other tubers (oca, lisa), grains (wheat, quinoa), and meat (guinea pig, sheep, pig, llama) varies by altitude and degree of market involvement. Farmers have long nurtured different varieties of potato and maize as a means of reducing risk and for different culinary functions. Some varieties of potato may be cultivated for commercial sale, but can be more susceptible to frost damage or late blight (*Phytophthora infestans*); others may be lower-yielding but more resistant to disease or to the effects of hailstorms (which can shred leaves). At higher altitudes, "bitter" varieties of potato are used to make *chuño*, a dehydrated (freeze-dried) form that can be stored for years. Farmers consequently intercrop these different varieties and segregate them for different uses on harvesting.

Although maize may not display a comparable level of genetic diversity in Bolivia to that found in Mexico – nor is it attributed the same kind of sacred value – there are nevertheless a very large number of varieties, distinguished largely on the basis of the colour of individual kernels, each of which performs different culinary functions. In the village of Chinchiri, Department of Cochabamba, where fieldwork was conducted during the 1980s, households cultivated the following varieties, among others:

- *Ch'eji* – large, grey-speckled kernels that were either toasted (*tostado*) or boiled (*mot'i*) and eaten as a midday meal in the fields;
- *Kulli* – dark purple kernels that were used for *chicha* (maize beer) or for *api*, a hot gruel drink (never used for eating);
- *Ch'uspillo* – a burnt orange appearance used for *tostado* and *chicha*;
- *T'una wilkapayro* – yellow and grey kernels suitable for most dishes, including *lawa*, a soup featuring potato, and *humintas* (fresh kernels are ground then made into a dough and baked in leaves);
- *Blanco* – large white kernels that are best for *mot'i* especially when peeled of their skin and served with *chicharron* (fried pork);
- *Amarillo* – yellow maize, suitable for most purposes (this variety and Blanco have Spanish as opposed to Quechua names, suggesting that they may be more recent, "improved" introductions).

The process of agricultural intensification, as outlined in chapter two, has had significant repercussions for biological diversity both within and outside cultivated systems. Globally, the expansion of agricultural land has been a major driver of destruction of moist tropical forest habitats, regarded as the most biologically diverse and productive biome. In the temperate North, the post-war shift from mixed farms to specialised intensive operations, and the accompanying search for efficiencies of scale, resulted in an expansion in the size of fields as farming operations proceeded through the different stages of motomechanisation. This led to the removal of hedgerows (lost at a rate of 18,000 km per year in the UK during the 1980s and 1990s according to Pretty 2002), and to the drainage of wetlands and seasonally flooded meadows. The loss of ground cover, nesting sites and hedgerow foods, including insects, has a significant impact on small mammals and especially birds, which are considered an indicator of environmental health. There was an especially startling decline in the population of farmland birds during the 1970s and 1980s in Britain, with large percentage declines across a number of species including the corn bunting, grey partridge, tree sparrow, song thrush and skylark.

Within cultivated systems, uniformity has displaced diversity as efforts have been directed at maximising crop productivity and food output. This has resulted in the widespread displacement of native landraces by high-yielding modern varieties, bred to be fertiliser-responsive and provided with protection from pests and disease by a host of agri-chemicals. Today, according to the MA, 80 per cent of the area in the South sown to wheat is under modern varieties, while three-quarters of the area under rice is sown to improved semi-dwarf varieties. Although this has undoubtedly increased aggregate output of food, there are growing concerns about the potential loss of plant germplasm as native landraces are displaced, and loss of the knowledge associated with them (their characteristics and attributes). A similar tale of genetic erosion can also be extended to domesticated animals, as a few highly productive breeds that have been selected for their ability to convert feeds to meat or milk are disseminated worldwide, displacing local varieties and resulting in their eventual extinction. Despite the efforts of local breeders, livestock – like varieties of apples or potatoes – are subject to commercial decisions that are made a long way from the farm.

Maintaining such diversity is vital, as this material might be considered the basic building blocks from which plant scientists draw in the future in developing new varieties capable of flourishing in a changing and warming world. Unfortunately, plant breeding has become dominated by some of the largest and most powerful agri-chemical companies, which have been systematically exploiting intellectual property right laws to appropriate and control plant germplasm. For example, the top five transnational seed companies control 47 per cent of global commercial seed sales and are seeking to extend their domination through the development and dissemination of transgenic seeds. However, other actors in the food chain also bear some responsibility for the narrowing of crop diversity. Retailers, for example, select varieties of fresh produce that offer longer shelf life and can be delivered blemish-free, while processors buying and shipping globally and in large volumes prefer uniformity for ease of purpose. The creation of international standards and regulations also contributes to this narrowing in the range of plants and animals.

It is becoming increasingly urgent to find an effective political–legal framework through which to properly acknowledge and compensate the vital role of farmers as custodians of this genetic reservoir, and those who endeavour to sustain the cultural knowledge of which biodiversity is an integral part (including the preparation of traditional foods). The genetic diversity of plants and animals, and the diverse knowledge and practices of rural communities, are two vital resources for adapting agriculture to the changing environmental circumstances that the world faces in the years ahead. Fortunately, there is growing evidence to indicate a widening understanding of this issue and willingness by many in North and South to take responsibility as savers of seed and guardians of this biological heritage.

Energy

If ecosystem services are essential to the maintenance of life on Earth, then it is the Sun and solar energy that underpins the performance of these services. The Sun:

● Lights and warms the Earth. The natural greenhouse effect traps some of the reflected radiation from the Earth's surface in the atmosphere and keeps the average ambient temperature at a comfortable 18°C. Without this, Earth would be nearly as cold as Mars.

● Drives photosynthesis, the capacity of green plants to absorb solar energy and to use this to synthesise carbon dioxide (as a source of carbon) and water (as a source of hydrogen) into carbohydrate.

The formation of plant tissue – leaves, stems, roots, fruit, seeds – through this process becomes the basis of food chains supporting many different kinds of life, from micro-organisms to grazing animals and humans.

● Drives the climate system, and through this regional and local weather patterns, responsible for the distribution of water (via the hydrological cycle that evaporates and precipitates freshwater) and heat around the Earth.

Although a surprisingly small amount (*c.* 0.8 per cent) of the total radiation from the Sun is used in photosynthesis, solar energy nevertheless dominates energy inputs into food crop production (*c.* 90 per cent), even in highly intensive systems. Despite extraordinary efforts in plant breeding, the efficiency of converting solar energy into plant tissue has not been substantially improved since the beginning of agriculture. As Mannion (1995) observes, it is sobering to acknowledge that the high-density, high energy-consuming industrial societies around the world are almost totally sustained by this comparatively small amount of solar radiation that, captured by vegetation aeons ago, now provides all our present-day fossil fuel needs.

In order to obtain food, humans intervene in and manipulate natural ecosystems. The greater the degree of change to the structure and composition of the ecosystem, the more energy is required to establish and maintain it. The productivity of the agricultural ecosystem – the agro-ecosystem – is limited by just the same factors that limit the growth of naturally occurring plants, that is, sunlight, nutrients, water, temperature and pests. By ensuring that these factors do not impose constraints on crop growth, farmers engage in tasks to ensure the provision of fertilisers, irrigation and pesticides. These inputs, as well as their provision, embody energy which is additional to that provided by the Sun; the greater the level of inputs and performance of field tasks (e.g. weeding), the greater is the external supply of energy (Box 3.5). Consequently, changing an ecosystem to a row-crop monoculture (whether of maize or potatoes) requires a substantial investment of energy in preparing the ground, sowing the seed, fertilising, preventing the invasion of early successional plants (weeds) and pests, and harvesting.

It is as well to remember that, throughout most of human history, the greater proportion of food production was performed by human labour. Yet producing crops by hand requires about 1200 hours per hectare (or 120 days @ 10 hours/day) according to Pimentel and Pimentel

Box 3.5

Energy analysis

Food is essentially a fuel for human beings: we need to eat to live. A certain minimum amount of food is needed to ensure basal metabolic function, and the more active the person, the greater the need for food to deliver energy output. Energy is the capacity to do work, and the use and conversion of energy is governed by the laws of thermodynamics. The first law states that energy cannot be created or destroyed, but can be transformed from one type to another (for example, solar energy is converted by photosynthesis to plant food energy). The second law of thermodynamics is that the transformation of energy can occur only from a concentrated to a more dispersed form (for example, an internal combustion engine converts the potential energy of petroleum into desired mechanical energy but with significant waste as heat loss to the environment).

The great value of energy analysis is that it can reduce quite different materials, such as food and oil, to their embodied energy content, thus we can evaluate the efficiency of conversion from one form to another. As solar energy is utterly indispensable to plant growth, it is not included in the calculations made in evaluating the conversion of human, animal, mechanical and other embodied energy inputs into food output. The energy ratio (E_r) is calculated as:

$$\text{energy out/energy in}$$

Energy analysis uses a number of different units, but in order to simplify matters, in this book we use the kilocalorie (kcal).

(2008), suggesting that each person can manage only one hectare during the growing season. Although it will vary by climate, in general terms under these conditions only the bare minimum of essential human needs can be obtained. Consequently, the utilisation of animal power to cultivate the soil was an immense breakthrough in agricultural production, with one hour of ox power substituting for three to five hours of human labour in preparing ground for planting. The use of horses was a significant improvement over oxen because they move more quickly. Although animals must be fed, most of their needs can be met through grazing outside cultivated areas. Yet, while animals substantially improved the efficiency of ploughing and transporting, the development of oil-powered tractors was to achieve an even greater facility to perform such tasks – though at a much greater cost in terms of energy input (see Box 3.6).

Box 3.6

A comparison of energy inputs to land preparation

A person using a heavy hoe to till 1 ha of soil for planting needs about 400 hours, or 40 days @ 10 h, to complete the task. Assuming that an individual expends 400 kcal/h for this heavy work, and with additional energy to maintain the worker for the remaining 14 h each day, they require 4850 kcal/day. In order to complete the **40 days** of work, the total energy expenditure is about 194,000 kcal. If an additional 6000 kcal is added for the construction of the hoe, this makes the total input needed to till 1 ha by human labour about **200,000 kcal**.

A pair of oxen take only **65 h**, expending an estimated 260,000 kcal, but require a person to drive them (estimated at 31,525 kcal) as well as the machinery input (6000 kcal). Thus oxen can prepare the ground using a little under **300,000 kcal**.

A small, 6 horsepower (hp) hand tractor would complete the task in **25 h**, using an estimated 23.5 litres of petroleum, equal to 237,000 kcal; with the amount of energy embodied in the machine and the human labour to drive it, this makes a total of **441,000 kcal**.

Finally, a 50 hp tractor, which is modestly powered by today's standards, is able to till 1 ha in **4 h**, using over 30 litres of gasoline and, together, with the other calculated inputs, would mean expending over **550,000 kcal**.

In other words, this tractor can perform the same task in just 1 per cent of the time required by human labour, but requires almost three times the energy needed for that worker over 40 days.

Source: adapted from Pimentel and Pimentel (2008), 60–1.

The development of high-energy inputs into food production grew dramatically in the post-war period, at least up to the first oil-price shock of 1973. Gerald Leach was one of the first to focus upon the declining energetic efficiency of the agri-food system, highlighting the growing reliance upon oil when it was costing $1.50 per barrel during the period 1950–70. As he notes, it was the equivalent of having a virtual human "energy slave" working for 4000 hours for a dollar. No wonder, he wrote, "that the cost calculations of farming almost everywhere stressed the need to increase yields and reduce manpower whatever the consequences for energy consumption through extra fertilisation, mechanisation, automation and the like" (Leach 1976: 2).

It was arguably the experience of cheap oil during this period that has so significantly shaped contemporary farming practice.

David and Marcia Pimentel (2008) have brought together a range of energy analyses of different farming systems in the USA and elsewhere. This comparative analysis provides some revealing insights into the relative energetic efficiency of widely different food production systems. Table 3.3 illustrates the primary inputs into a highly mechanised modern farming system, maize (corn) production in the United States. The table lists each principal component, with quantities of each input and its respective embodied energy (in kcal) per hectare. What is most apparent is the modest level of human labour, mostly spent in driving machinery, and the high level of fossil energy inputs derived from petroleum and natural gas. The consumption of diesel and gasoline (petrol) accounts for one-quarter, while the single largest input is represented by nitrogen fertiliser, manufactured from natural gas, which constitutes 30 per cent of the total energy input. According to the Pimentels, in US corn production around one-quarter of the energy inputs are used to reduce human and animal labour inputs, and the remaining three-quarters to increase yield. Overall, the energy ratio (E_r), energy out/energy in, is 3.84.

Table 3.3 Energy inputs in US maize (corn) production

	Quantity (per ha)	Energy ('000 kcal/ha)
Inputs		
Human labour	11.4 h	4.65
Machinery	55 kg	1018
Diesel	88 litres	1003
Gasoline	40 litres	405
Nitrogen	153 kg	2448
Phosphorus	65 kg	270
Potassium	77 kg	251
Limestone	1120 kg	315
Seeds	21 kg	520
Irrigation	8.1 cm	320
Insecticides	2.8 kg	280
Herbicides	6.2 kg	620
Electricity	13.2 kWh	34
Transportation	204 kg	169
Total		8115
Outputs		
Corn yield	8655 kg	31,158
kcal output/kcal input		**3.84 : 1**

Source: adapted from Pimentel and Pimentel (2008): 105.

The utility of energy analysis is that it can provide an entirely fresh perspective, and can challenge many of the usual assumptions associated with the pursuit of productivity. When we begin to appreciate that many of the simplest farming systems involving polycultures (mixed, densely planted gardens) achieved some of the highest energy ratios (well in excess of 10:1), then there may be reason to re-examine their attributes, especially in light of the anticipated tightening of energy supplies, to be discussed in chapter four. On the other hand, some food production systems are extremely energy inefficient, with capture fisheries and aquaculture performing least well.

With regard to capture fisheries, evidence indicates that small-scale and coastal fishing is relatively energy efficient, achieving parity ($E_r = 1$) in terms of energy expenditure to fish protein harvested, while also providing employment and lower fuel costs. Unfortunately, the drive to scaling-up, with larger vessels travelling greater distances to capture more fish, has seen energy ratios deteriorate quickly beyond 20 kcal of fossil energy input for every kcal of fish protein output ($E_r = 0.05$). Moreover, most of the top ten marine fish species, which together account for around 30 per cent of all capture fisheries' production, are considered fully or overexploited. Yet aquaculture does not achieve better efficiencies or environmental performance. Freshwater catfish production in Louisiana requires 34 kcal of fossil energy for every kcal of catfish protein, which is comparable with US beef production. The farming of Malaysian prawn in Hawaii achieves a ratio twice as bad as that of catfish, 67:1, yet remains commercially profitable – if energetically profligate – because of the high market value of prawn. In contrast, the practice of farming salmon in cages along the Norwegian coast achieves an energy ratio between the two, 50:1, in which much of the energy subsidy is used to fish down the food chain and harvest low-value species over large areas of ocean, and convert this to pellet feed for the caged salmon.

Energy analysis places in sharp relief the capacity of the industrial agri-food system to discount the extraordinary subsidy provided by fossil fuels in order to produce, in abundance, a wide range of foods. It reveals that for every 1 kcal of salmon protein that we might wish to consume, 50 kcal have been expended in producing this salmon. Note that this accounts only for its production, and does not take account of the embedded energy in getting this fish into the supermarket, let alone onto our plates, a topic addressed in chapter five. The situation for livestock products is comparable, with beef and lamb at the upper end and poultry meat at the lower end of the energy ratio scale. The livestock sector is discussed in chapter four.

In summary, energy analysis has highlighted how the use of highly powered machinery to perform field tasks quickly may have made economic sense when there was an apparent abundance of petroleum at low price and where rural labour could be redeployed in industrial jobs. It may be the case, however, that we are entering into a new phase of human history, where the hidden subsidy of "slave labour" represented by cheap fossil fuels is likely to no longer be available, and as energy costs rise, farming systems might require higher levels of investment of human labour power once more.

5. Summary

This chapter began by reviewing the vital role performed by ecosystem services to primary food production, noting how the majority are under significant pressure as a result of agricultural intensification. Although it was suggested that the availability of land need not be the primary constraining variable in realising sufficient output to feed 10 billion people, the composition of our diet may need to reflect better the urgency of stemming the rate of land-use conversion, which is driven partly by the expansion of the area under livestock feeds.

Following a brief elaboration of different farming systems that helped to establish the key variables of moisture, temperature and patterns of plant and animal domestication across the Earth's surface, the chapter then proceeded to examine four key resources for primary food production. It is clear that the stock of soil, water and biological resources is vital to maintaining sufficient levels of food output into the future, yet an attitude of utilitarian complacency amongst those managing the most intensive farming systems has left such resources in a vulnerable state. One of the keys to understanding the drive toward mechanised intensification is the role played by cheap fossil fuels, especially natural gas, in the manufacture of synthetic fertilisers; and by oil in driving farm machinery. Such contributions are like having a virtual energy slave, permitting a huge substitution of human labour by chemical and mechanical aids.

Unfortunately, this model of food production may no longer be feasible, as we look into a future of rising energy prices. Chapter four examines the phenomenon of "peak oil", a concept that is gaining increasing acceptance amongst energy analysts worldwide and that is likely to have enormous implications for primary food production. Other challenges that threaten to have a significant impact upon the prevailing paradigm of primary production include climate change and increasing scarcity of freshwater resources. Taken together, it may be necessary for all of us who enjoy dietary choice to look more closely at the composition of our food intake, most especially the rising global appetite for meat and other livestock products.

Further reading

Kimbrell, A. (ed.) (2002) *Fatal Harvest: The Tragedy of Industrial Agriculture*. Sausalito, CA: Foundation for Deep Ecology/Island Press.
This is a large-format, highly impassioned text documenting the destruction of the rural environment and family farm economy in the United States as a result of industrial agriculture. It contains around fifty essays and some superb photographs.

Kurlansky, M. (1999) *Cod: A Biography of the Fish that Changed the World.* London: Vintage Books.
The decision to focus on terrestrial food production means that my book deliberately neglects capture fisheries. But one book that more than compensates is Kurlansky's account of one fish species through history.

Pimentel, D., Pimentel, M. (2008) *Food, Energy and Society*, 3rd edn. Boca Raton, FL: CRC Press.
This book comprises more than twenty chapters, each examining aspects of food production, authored or co-authored by David Pimentel, a highly distinguished professor of ecology and agricultural science at Cornell University. A vital body of work.

And on the web:

Millennium Ecosystem Assessment: www.maweb.org/en/Reports.aspx

The MA did not undertake primary research, rather it brought together a very wide range of existing information and sought to evaluate and communicate it in a way that would demonstrate the serious state of disrepair of ecosystem services. The exercise resulted in the publication of a huge volume of material, some parts more accessible and comprehensible than others. Although many of the reports can be purchased as hard copies, there is some value in browsing through the online materials, particularly the Synthesis Reports.

4 Global challenges for food production

Contents

1. **Introduction**
2. **Climate change**
   Effects of climate change
   Agricultural greenhouse gases
3. **Global freshwater resources**
   Extracting water
   Virtual water
   Water footprints
4. **Peak oil and its implications**
   Future projections
   Biofuels
5. **Livestock and rising demand for meat**
   Meat consumption
   Livestock systems
   Factory farming
   Environmental impacts
6. **Summary**
Further reading

Boxes

4.1 Carbon sequestration options
4.2 Mining the Ogallala aquifer
4.3 Damming the rivers
4.4 Draining the Aral Sea
4.5 Canada's tar sands: dirty oil
4.6 Bovine culture
4.7 Feedlot beef
4.8 A pig's life

Figures

4.1 Greenhouse gas emissions from the UK food chain, 2006
4.2 Sources of agricultural greenhouse gas emissions
4.3 Share of global field-level water use by different crops
4.4 Virtual water flows: leading exporters and importers
4.5 Discovery and production of regular oil
4.6 The depletion curve for oil and gas
4.7 World annual fuel ethanol production 1975–2009

Tables

4.1 Global average virtual water content of selected primary foods
4.2 Meat consumption per capita by region
4.3 Population of the principal commercial livestock categories by world region

1. Introduction

Chapter three reviews the ecological basis for primary food production, providing a brief characterisation of the principal farming and other primary food production systems, and outlines the significance of four key resources that underpin agriculture in particular. Without specifying limits or thresholds, it is clear that, through neglect or inappropriate development, there is a danger that vital ecosystem services could be lost unless efforts are made to restore farming methods that provide better equilibrium with available resources. In this chapter, we consider four major challenges that especially highlight this disparity.

Climate change, freshwater depletion, peak oil and rising demand for livestock products are quite separate issues that need to be dealt with in turn, and in some depth, in order to make clear why they have been singled out for particular attention. Yet it is also apparent that each does not exist entirely independently of the others. In addressing climate change, for example, we need to recognise how the rapid and, arguably, heedless exploitation of fossil fuels has resulted in atmospheric concentrations of greenhouse gases that are responsible for global warming, while at the same time presenting us with the predicament of "peak oil". A warming world, for many parts of the globe, will also probably become a drier one, putting greater strain on local water resources and presenting severe challenges for production systems, whether for local subsistence or distant markets. In this regard, our dietary choices, too, have consequences for our ability to reduce the volume of greenhouse gas emissions, and will influence whether we will be successful in stabilising the global climate system. Consequently, although the primary objective of this chapter is to drill down into each issue and reveal the scale and complexity of the challenge that each presents, the reader is reminded to keep in mind the range of interconnections between them – as well as other global predicaments.

2. Climate change

As we have seen, primary food production exerts many different kinds of environmental impact, from land-use change, through the consumption of resource inputs, to the generation of agricultural wastes.

Most of the activities associated with primary food production are also a source of greenhouse gas emissions. This section reviews the challenge that climate change poses for food production, then goes on to examine the contribution that agriculture makes to global warming. It is important to recognise that all components of the agri-food system, from the agri-technologies supply industry, through primary production, processing, manufacturing, distribution and retail, right down to the domestic arena of the storage, cooking and disposal of food, are responsible for the emission of greenhouse gases. One estimate for the entire EU food supply chain suggests that what we eat has more impact on climate change than any other aspect of daily life, accounting for 31 per cent of the global warming potential of products consumed within the then EU-15 (Tukker *et al.* 2006). This section concentrates on agriculture's contribution to climate change and how it will be affected by it; chapter five looks in more detail at the stages beyond the farm gate.

The contribution of agriculture alone varies significantly between countries, reflecting the mix of different farming systems. For example, in Ireland, with its large dairy industry and other livestock activities, agriculture accounts for 28 per cent of the country's total greenhouse gas emissions, whereas for the EU as a whole agriculture's contribution is around 10 per cent. In the UK, total greenhouse gas emissions from the food chain were estimated to be around 160 mt of CO_2-equivalent (see below for an explanation of this), representing around 22 per cent of total UK economic activity (Defra 2009). Figure 4.1 provides a visual breakdown of the UK food chain. It shows that primary food production – comprising farming and fishing activities within the UK – account for 53 mt of CO_2-equivalent emissions, or just over one-third of the total for the entire food chain. Next by importance, amounting to 39 mt of CO_2-equivalent (around a quarter of total emissions), is the net balance of food imports over exports, which relate exclusively to production operations in countries outside the UK. The third largest source of food-related greenhouse gas emissions (21 mt) arises from UK households in their performance of shopping, storage and preparation of food. It is worth noting that the relative share of this category has grown from 11 per cent in 2002 to 13 per cent of total emissions in 2006, with almost half of this increase due to food shopping by car (Defra 2009). By comparison, some of the other elements of the food chain appear to be modest contributors of greenhouse gases. The environmental impacts of these elements are discussed in chapter five.

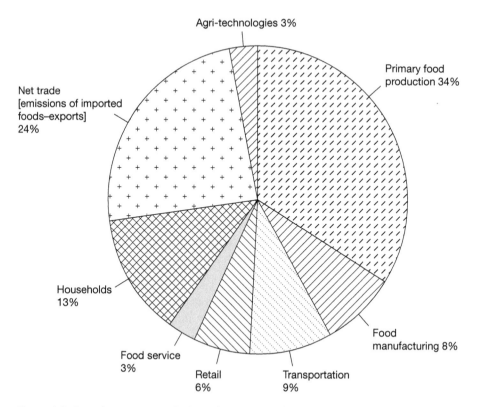

Figure 4.1 *Greenhouse gas emissions from the UK food chain, 2006*
Source: adapted from Defra 2009: 46.

Effects of climate change

At the beginning of the section on energy in chapter three, it was noted that the natural greenhouse effect is a positive feature of life on Earth; however, the *enhanced* greenhouse effect is a consequence of the increased atmospheric concentrations of greenhouse gases. These gases, principally carbon dioxide (CO_2), methane (CH_4) and nitrous oxide (N_2O), absorb the reflected radiation from the Earth's surface and are warmed by it. As their atmospheric concentrations have grown – in the case of CO_2 from 280 parts per million by volume (ppmv) in the pre-industrial era to 390 ppmv today – so the atmosphere has become warmer: average global temperatures have risen by 0.76°C in the past 100 years. The phenomenon of human-induced climate change with warming of the climate system is now regarded as unequivocal.

According to the evidence, the twelve warmest years on record within the past 150 years have occurred during the past thirteen years. There are particular anxieties in relation to the following.

● The rate of loss of the polar ice sheets – recent data from ongoing monitoring indicate that the rate of loss is proceeding more quickly than was predicted by the Intergovernmental Panel on Climate Change Fourth Assessment Report (AR4) (IPCC 2007). The consequences of ice sheet retreat include a reduction in the level of reflectivity so that more heat is absorbed by the Earth's surface, particularly the oceans. Moreover, the volume of meltwater will contribute to rising sea levels at a much greater rate than anticipated by AR4, with studies since 2007 ranging from a 0.5 to 2 m rise by 2100 (Pew Centre 2009).

● Tipping points – rather than a continued linear response between warming and ecological effects, scientists are concerned about critical thresholds that could trigger accelerated rates of change. The loss of Arctic sea ice and the retreat of the Greenland ice sheet are particular causes of concern, as is the rate of thawing of permafrost in the Arctic Circle that would release huge quantities of methane.

● An increase in the frequency and intensity of extreme events related to meteorological phenomena such as hurricanes, tropical cyclones, heatwaves, droughts and floods. A recent report draws on data from the global reinsurance group Munich Re, and notes that "in 2007 there were 960 major natural disasters (the highest ever such figure) with more than 90% being the result of extreme weather-related, or climate-related events, together accounting for 95% of the 16,000 reported fatalities and 80% of the total $82 billion economic losses" (Lancet 2009: 1706).

Of course, climate change will not be felt equally everywhere, and the consequences for primary food production will also vary significantly. For example, in mid- to high-latitude regions of the world (North America, Russia, Central Asia, Northern Europe), it is thought that the rise in ground-level temperatures will lengthen the growing season and should increase yields, at least in the medium term. Increased levels of atmospheric CO_2 will also have a generally positive effect on crop growth by enhancing photosynthesis. During the next two decades, it may be possible for these regions to grow commercially crops that have not been viable in the past, such as wine grapes and other temperature-sensitive fruits, as well as to extend the northern limit of cereal cultivation. Yet, as warming proceeds, there will be increased evapotranspiration and soil moisture losses that will be exacerbated in many regions by a projected fall in precipitation levels. By mid-century, excessively high temperatures and water shortages, as well as a likelihood of increased plant disease, fungal infections and pest outbreaks, will threaten crop yields.

For the developing regions of the low latitudes (within the tropics), climate change brings not only warming, which is a concern where crops may already be close to their limits for heat and water stress, but greater variability and more erratic rainfall. Projections of changing precipitation patterns will particularly affect rainfed agriculture – which covers 96 per cent of all cultivated land in Sub-Saharan Africa, 87 per cent in South America and 61 per cent in Asia. This will present an enormous challenge in regions where there is already considerable vulnerability, where the majority of households depend most on agriculture and where there are fewer alternative sources of income, and consequently more limited adaptive capacity. With climate change also responsible for changes in the frequency and magnitude of more extreme events – droughts, storms, flooding – it is apparent that in developing regions there is increased risk of significant impacts on food production systems. Potentially, the biggest losers from climate change are likely to be people most exposed to the worst of its impacts and the least able to cope while, cruelly, living in countries that have made the least contribution to anthropogenic warming.

Agricultural greenhouse gases

But what is the contribution of primary food production to climate change? As noted earlier, there are three principal greenhouse gases: carbon dioxide (CO_2), methane (CH_4) and nitrous oxide (N_2O), as well as a number of industrial chemicals (some of the fluorocarbon family used in refrigeration) that are long-lived with high global warming potential. Although CO_2 is the single most important greenhouse gas by volume and is responsible for about 85 per cent of the UK's contribution to global warming, agricultural production is more closely associated with emissions of methane and nitrous oxide. These gases are more powerful than CO_2, having a global warming potential twenty-three times that of CO_2 in the case of CH_4, and 296 times for N_2O. While CO_2 is most closely associated with the burning of fossil fuels, CH_4 is closely linked to livestock, especially ruminants that possess four gastric chambers through which they are able to break down cellulosic plant material. This digestive system generates methane, which is released at either end of the animal. Wetland rice cultivation is also a source of methane. N_2O emissions, on the other hand, derive from the volatilisation of nitrogen from the soil, principally that derived from chemical fertilisers. Much research has suggested that crops may not take up even half the nitrogen made available through the application of fertiliser; the remainder is either leached through the soil creating pathways of nitrate contamination in ground and surface water, or is released as N_2O into the atmosphere.

Drawing on data in the AR4 of Working Group III of the IPCC (Smith *et al.* 2007), it is possible to consider the contribution of agriculture to the emission of greenhouse gases. It notes that in 2005, agriculture accounted for between 5.1 and 6.1 Gt CO_2-eq per year (i.e. 5.1 billion tonnes of CO_2-equivalent per year), which represents between 10 and 12 per cent of global anthropogenic emissions of greenhouse gases (those derived from human activities, not natural ecosystem sources). Of this, CH_4 contributed around 3.3 Gt CO_2-eq per year, and N_2O 2.8 Gt CO_2-eq per year, which represents 50 and 60 per cent, respectively, of the total global anthropogenic emissions of these gases. On the other hand, agriculture is responsible for only a relatively small net contribution of CO_2 (in the order of 40 million tonnes) despite the large annual exchanges between the atmosphere and terrestrial production systems. This excludes all post-farm-gate transportation and processing activities. The individual sources of global agricultural greenhouse gas emissions are shown in millions of tonnes of CO_2-equivalent in Figure 4.2. It must be remembered that emissions from soils and from cattle stand out due to the higher global warming potential of N_2O and CH_4.

However, as far as the two primarily agricultural greenhouse gases are concerned, evidence from the recent past and projections into the short term rather cloud optimistic prospects for the medium to long term. Over the period 1990–2005, global agricultural emissions of CH_4 and N_2O grew by 17 per cent, or by 60 mt CO_2-eq per year. This growth was most marked in those regions of the global South and economies in transition, so that by 2005 they collectively accounted for three-quarters of total global agricultural emissions. The regions of the developed North, in contrast, displayed a decrease in the emissions of these gases, led by Western Europe, which has been implementing a range of agri-environmental measures. Forward projections of global agricultural emissions to 2030 are not encouraging, however, with N_2O emissions

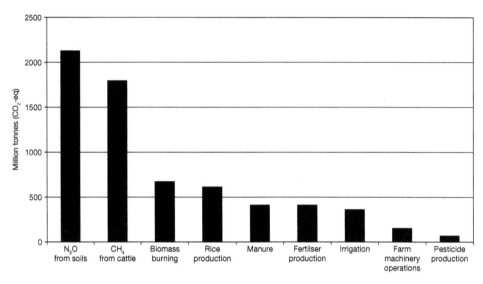

Figure 4.2 *Sources of agricultural greenhouse gas emissions*
Source: drawn from data presented in Bellarby et al. 2008.

expected to increase by between 35 and 60 per cent due to more widespread use of chemical fertilisers, and CH_4 by 60 per cent in direct proportion to an anticipated increase in livestock numbers.

In this regard, the challenge of mitigating climate change through reduced emissions of these two greenhouse gases most closely linked to food production becomes apparent. With rice such an important staple crop in East and South-East Asia, as well as elsewhere, a projected increase in irrigated area is likely to lead to a 16 per cent increase in CH_4 from paddy cultivation. Meanwhile, the Middle East and Africa have the highest projected growth in agricultural emissions: a 95 per cent increase from 1990 to 2020, much of which will be attributed to N_2O. Given the need to enhance food production capacity in Sub-Saharan Africa, which has been marked by poor and declining levels of soil fertility and low levels of fertiliser use, it is understandable that efforts would be made to increase rates of fertiliser application in pursuit of improved food security. Meanwhile, in Latin America and the Caribbean, significant changes in land use and land cover, with forests such those as in Amazonia converted to crop- and grassland and with a resulting doubling of cattle numbers between 1961 and 2004, there have been, and will continue to be, increases in emissions of CO_2 and CH_4.

As for climate change-mitigation options, the agricultural sector offers some of the more technologically and commercially feasible opportunities with the potential to change the position of primary food production from the second largest source of greenhouse gas emissions to a much lesser contributor, even a net sink. Although efforts should continue to be directed toward ways of reducing CH_4 and N_2O, there are opportunities to sequester carbon through improved land and soil management practices. Some options are outlined in Box 4.1. There is also some enthusiasm in certain quarters for the development of *biochar*. Subjecting particular kinds of biomass to industrial processes such as pyrolysis gives rise to gases that can be liquefied and used as fuels for energy generation, with the residue comprising a highly porous form of charcoal that, if added as an amendment to soils, is believed to enhance fertility and to sequester significant quantities of carbon for hundreds of thousands of years. It might indeed be considered as the magic bullet of carbon capture and storage for the land-based sector's contribution to anthropogenic climate change (Biochar Research Centre 2009).

Box 4.1

Carbon sequestration options

1 *Cropland management* – mitigation potential up to 1.45 gigatonnes of CO_2-equivalent per year (Gt CO_2-eq/year):

- avoiding bare soil through use of catch- and cover-crops between the main commercial crops, thus preventing erosion and leaching of nutrients, and locking up carbon;
- "smarter" applications of nitrogen fertiliser, for example in smaller quantities and at the right time for optimum uptake by crops, and making greater use of nitrogen-fixing legumes to compensate for less fertiliser use;
- eliminating the burning of crop residues (such as corn stubble);
- reduced tillage – no-till farming avoids ploughing, and minimal soil disturbance optimises carbon sequestration.

2 *Grazing land management* – mitigation potential up to 1.35 Gt CO_2-eq/year:

- reduced grazing intensity (lower livestock numbers) and use of fire (to burn off noxious plant species) can enhance CO_2 uptake in the soil and in biomass.

3 *Restoration of degraded lands and drained wetlands* – mitigation potential up to 2 Gt CO_2-eq/year:

- adding nutrients and organic amendments to degraded soils can restore their productivity and carbon-holding capacity, to avoid the further conversion of wetlands and to restore those no longer suitable for crop production.

4 *Improved water management* – mitigation potential up to 0.3 Gt CO_2-eq/year, particularly in wetland rice farming.

5 *Set-aside and improved manure management* – mitigation potential 0.3 Gt CO_2-eq/year, including converting cropland to agro-forestry.

6 *Improved efficiency in manufacturing fertilisers* – mitigation potential up to 0.2 Gt CO_2-eq/year: energy-saving and improved nitrous oxide-reduction technology.

Source: adapted from Bellarby *et al.* (2008).

3. Global freshwater resources

In chapter three, within the evaluation of resources required for primary production, the utter indispensability of water was made clear. Yet there is a very uneven distribution of freshwater resources across the Earth and, in some regions, the shortfall is becoming more acute not only for food production, but for meeting basic human needs. It has been calculated that agriculture accounted for 90 per cent of global freshwater consumption during the past century. Yet the global population is projected to increase to more than 9 billion by 2050, while the international community struggles to make progress on the Millennium Development Goals' poverty targets (that prioritise meeting the water and sanitation needs of the poorest), and all this at a time when the global climate system is entering a warmer, more unstable phase. This will inevitably place further stress on water resources (Khan and Hanjra 2008).

This section reviews the challenges posed by the utilisation of water for food production, highlighting some of the problems arising from a paradigm that appears to have treated freshwater as a limitless resource. Recent years have, however, seen the emergence of new and important conceptual innovations that reveal the short-sightedness of pre-existing attitudes to water as an infinite gift of nature, and indicate possible pathways toward a more rational and sustainable utilisation. While outlining some of these new concepts, it will nevertheless become clear that without some significant changes to the underlying rationale of international trade, the potential crisis facing global freshwater resources will be realised.

Extracting water

Prior to the modern, post-1945 era, much of the world's food needs, at least beyond the leading industrialised countries, were met through national or subnational provision, most commonly at the scale of the river basin, with the moisture needs of agriculture supplied by rainfall or through the diversion of surface water resources. In the latter case, the vast majority of irrigation schemes largely comprised simple, gravity-fed systems that drew from streams and rivers, and involved the channelling of water along furrows, or the flooding of fields as in paddy rice cultivation. Rice has been grown under irrigation in the Far East for 5000 years, with rice terraces in Banaue, the Philippines under continuous irrigated cultivation for 3000 years, illustrating the sustainable nature of these maintained structures. Although gravity-fed systems are generally technologically simple and easily maintained, they are inefficient as up to half the water supplied can be lost through seepage, spillage and evaporation. In the absence of surface water resources at a higher altitude than the fields, native ingenuity found ways of lifting water, for example by using variations of Archimedes' screw and the Persian wheel powered by people and animals, or wind-powered pumps. However, it is with the development and widespread availability of pumps powered by diesel oil and electricity that the constraints of channelling surface water or shallow groundwater could now be overcome, and the resources of deep aquifers exploited. Box 4.2 provides an example of how a particularly large subterranean water resource can be seriously depleted by the widespread use of pumps supplying high-pressure sprays and sprinkler systems.

The harnessing and management of large hydrological resources does appear to offer enormous potential for agricultural and economic development. Indeed, it has been argued that the first agricultural civilisations – Ancient Egypt, Mesopotamia, China, India and pre-Columbian Peru and Mexico – were characterised by a high degree of social organisation based around irrigation and the management of water. Although the notion of hydraulic civilisations of Karl Wittfogel (1896–1988) has been subjected to profound critique from scholars, there is nevertheless a plausible argument to be made that certain states do indeed appear to be attracted by the social prestige and political capital to be gained from engaging in large-scale hydraulic engineering projects. Box 4.3 provides examples of giant dam construction and their consequences.

Box 4.2

Mining the Ogallala aquifer

The Ogallala aquifer is an example of a non-renewable stock resource rather than a renewable flow resource. It stretches under portions of eight states, from South Dakota to Texas, covering the High Plains of the USA, a semi-arid region marked by low (< 500 mm) and variable annual precipitation. Without irrigation, the High Plains ecozone would support extensive grazing, either of the native buffalo that once prevailed across this landscape, or the cattle that came to replace them. Initial efforts to develop arable farming resulted in the Dust Bowl years of the 1930s. Today, however, 94 per cent of the water withdrawn from the aquifer supports 5.5 million ha of irrigated crop production, largely of cereals, soybeans and cotton, and constitutes fully one-fifth of the total annual US agricultural output, worth more than $20 billion.

The Ogallala was created several million years ago by streams flowing from the Rocky Mountains, but geological activity has since cut this source of mountain run-off so that there is no longer any effective recharge of the aquifer. Systematic mining of the Ogallala began in the late 1940s and early 1950s, and the popular view was that the resource was inexhaustible. The number of wells soared (in west Texas alone from fewer than 1200 in 1937 to 66,000 in 1971), as did the rate of withdrawal (a fivefold increase between 1949 and 1974). In some areas, the rate of extraction was causing a 1.22–1.83-metre fall in the level of groundwater each year, while precipitation (rainfall and snow melt) was able to restore just 1.27 cm. In 2000, total groundwater withdrawal from the aquifer for irrigation amounted to 21 million acre-feet (or 26 km^3). Area-weighted, average water-level change from pre-development to 2005 shows a decline of 3.9 m, with some areas recording declines in excess of 100 feet (30.5 m). In southern Kansas, wells have dried up and farmers have had to revert to no-till dryland farming methods in order to grow wheat. Elsewhere, smart robotic irrigation systems are improving the efficiency of water use and reducing rates of extraction. However, new developments threaten to put the aquifer under increasing pressure: the expansion of the biofuels sector is a major economic incentive for growing more maize, while there are plans to tap the aquifer in order to supply drinking water to the cities of Dallas and even El Paso, over 1000 km away.

Sources: adapted from McGuire (2007); Braxton Little (2009).

Box 4.3

Damming the rivers

India lies third in the league of dam-building nations, and has been engaged in the highly controversial Narmada River scheme, most attention to which has focused on the Sardar Sarovar Dam. The height of the dam has been raised on several occasions since its initial formulation, and is now expected to reach 122 metres that will impound a reservoir 214 km in length and involve the displacement of 240,000 people. Proponents of the scheme argue, however, that the 66,000 km of canals will allow for the irrigation of 1,845,000 ha of low-productivity drylands.

China sits at the top of the international league of dam-building nations, with around half of the world's large dams. It has come under most criticism, however, for its Three Gorges scheme, the world's largest hydropower project, which sets records for number of people displaced (more than 1.2 million); number of cities and towns flooded (thirteen cities, 140 towns, 1350 villages); and length of reservoir (more than 600 km). Critics argue that the scheme is vulnerable to a number of environmental problems, ranging from the hazards of industrial pollutants and siltation in the reservoir, to the possibility of reservoir-induced seismicity. The latter notion suggests that earthquakes may be triggered by the extra water pressure created by a large reservoir, and that the construction of high dams in seismically active regions should be avoided. More geophysical research is under way to improve understanding of the propensity of regions to reservoir-induced seismicity.

Source: adapted from International Rivers (2009).

The construction of dams on rivers has been a focus of sustained criticism over many years. Indeed, there is an established catalogue of well documented social and environmental problems arising from large dams. These include: problems associated with the impounding of water and the creation of large, often quite stagnant lakes that occupy significant areas of arable land and displace previously resident farm families; the drowning of forests and the drainage into the reservoirs of nutrients which significantly raises the potential for eutrophication; and increased exposure of neighbouring populations to vector-borne diseases (such as malaria and schistosomiasis). These constitute some of the issues upstream of the dam, in the catchment zone. Below the structure, other problems arise associated with irrigation schemes themselves, particularly salinisation of the soil. This can occur in low-lying arid and semi-arid regions, where groundwater is often quite heavily charged with salts and irrigation causes waterlogging, the water table rises, and intense evaporation leaves salts in the uppermost layers of the soil. It has been estimated that around one-fifth of the world's irrigated land suffers from salinisation or waterlogging, with around 1.5 million ha per year lost from production. The dangers of inappropriate irrigation cannot be underestimated, as the final example in Box 4.4 illustrates.

Although large-scale dam building remains popular amongst technocrats, the net rate of increase in irrigated area has fallen steadily for the past

---

Box 4.4

Draining the Aral Sea

The Aral Sea lies in Central Asia, occupying territory in Kazakhstan and Uzbekistan, and in 1960 it was the fourth largest lake in the world. It has no outlet, but equilibrium existed between inflow, provided by two principal rivers, the Amu Darya and Syr Darya, and the evaporation rate of about 60 km^3 per year. During the 1920s, the Soviet Government initiated a programme of cotton cultivation in the region, and this was massively expanded during the 1950s. Under the control of collective farms, millions of acres of irrigated land were created, fed by large volumes of water diverted from the two rivers before they reached the Aral Sea. Though it still received an estimated 50 km^3/year of water in 1965, by the 1980s the flow of freshwater into the sea was zero. The Aral Sea consequently began to shrink, the shoreline retreated, and more than 28,000 km^2 of former lake bed were exposed. Strong winds picked up an estimated 43 million tonnes of sediment laced with salts and pesticides, and the resulting dust storms in the region are attributed with causing a huge increase in incidences of respiratory illness and cancers, especially of the throat.

As the Sea shrank – its surface area was reduced to half by the 1990s and its volume by 75 per cent – the concentration of salts and minerals in the water resulted in a dramatic alteration of the lake's ecology. Fish stocks collapsed, and with them the fishing industry, which had employed up to 60,000 people. By 2000, the Sea had split into two and the southern part had further split into western and eastern lobes.

The disappearance of the Aral Sea has also had an effect on the regional climate, its moderating influence giving way to shorter, hotter and drier summers and longer, colder winters.

A time-series of satellite photographs showing the retreat of the extent of the Aral Sea during the 2000s can be seen online at http://earthobservatory. nasa.gov/Features/WorldOfChange/aral_sea.php

Sources: adapted from Millennium Ecosystem Assessment (2005a); UNEP-GRID (2010).

four decades to less than 1 per cent per year as unexploited freshwater resources become more limited. Many countries within the band stretching from China through India and on through the Middle East to North Africa either currently or will soon experience insufficient water supply to maintain per capita food production from irrigated land. Moreover, agriculture is facing increasing competition for freshwater resources from other sectors: the needs of growing urban populations, industry, fisheries, recreational use and demands for the protection of natural ecosystems. It is estimated that water scarcity will affect 40 per cent of the global population by 2050, and with climate change scenarios suggesting reduced rainfall in some areas (and the likelihood of increased flooding in others), water stress is likely to grow.

Consequently, it is being suggested that we need to switch our approach from "humid zone thinking" (that encourages us to ask how much water we need and where we get it from) to one better suited to the rapidly emerging situation of water scarcity (how much water is there, and how can we optimise its use?) (Crabb 1996). Water *availability* is a function of nature and the hydrological cycle; water *scarcity* is a consequence of inappropriate human activity ill-suited to the resources in a particular location. A perspective that begins from an understanding of scarcity and the urgent need for more sustainable management of water resources will probably result in a variety of different responses.

Economists, for example, have promoted water charges as a means of regulating use, believing that a market price for water will emerge that will secure improved efficiencies in its use. But it is critical that operationalising such measures do not become a means of bolstering private profit over human need, as appeared to be the case in the disastrous efforts to privatise water supply in Cochabamba, Buenos Aires and Manila in the 1990s. Engineering agronomists accustomed to achieving a "technological fix" for resource constraints are meanwhile focused on devising new ways of improving the efficiency of water delivery and uptake. These include "high-tech" sensors that calculate crop needs, and robotic irrigation systems, as well as the genetic engineering of more drought-tolerant crops.

On the other hand, there are a huge number of "lower-tech" but tried-and-trusted measures that can be drawn upon. These range from drip irrigation systems that deliver water to plant roots; through the use of deficit irrigation, which delivers less water than the crop might optimally use, but where the modest decline in yield is more than compensated by savings of water; to well established water conservation methods such as low-tillage farming, mulching, rainwater harvesting or the use of wastewater in urban and peri-urban areas.

While welcoming "more crop per drop" technical innovations, this cannot distract us from more fundamental concerns about the underlying assumptions of water use and its continued availability in some regions of the world. It is clear there is a need for improved governance of water at a variety of different scales: from the global through the national level, to drainage basins and village committees, to individual water users. Above all, a more integrated approach is required in which a human right to safe drinking water and sanitation is established as the first priority of any scheme engaged in harnessing hydrological resources, whether for the generation of hydropower or the delivery of irrigation. Naturally, it also raises questions about the logic of a global agri-food system built upon the principle of comparative advantage, where countries and regions have become almost "designated" producers of certain commodities – reflecting their historical roles within the world economy and its international division of labour, the relative value of factors of production (especially land and labour), and domestic policies – rather than reflecting their actual endowments of water and other resources. This international trading system has consequently given rise to the cultivation of unsuitable dryland environments with inappropriate crops destined for distant markets, rather than prioritising the pursuit of local food and water security.

Virtual water

To speak of our appetite for "eating water" is to draw attention to the volumes of water locked up – or embedded – in primary foods, for which the term "virtual water" is now widely used (Roth and Warner 2008). Different crops have different water requirements, which will also vary according to the climates in which they are cultivated, and it is possible to calculate the virtual water content of all foods. The global average for maize, wheat and husked rice, for example, is 900, 1300 and 3000 m^3 of water per tonne, respectively, the higher figure for rice reflecting the levels of evaporation from flooded paddy fields as well as the weight loss arising from processing the grain (Hoekstra and Chapagain 2007). Livestock products have much higher virtual water content than crop products because of their extended production time and the nature of inputs required. For example, beef produced under industrial farming takes, on average, three years to reach the slaughter weight of 200 kg of boneless beef. During this time, the animal consumes nearly 1.3 t of grain, 7.2 t of roughage (pasture, silage), 24 m^3 of water for drinking, and 7 m^3 of water for servicing. By calculating the amount of virtual water within the feeds, it is possible to estimate the total volume of embodied water in the beef. Although this varies between countries, Hoekstra and Chapagain (2007) suggest a global average of 15,500 m^3 per tonne, or, in terms that we can grasp more easily, it takes around 15 m^3 of water to produce 1 kg of beef. The virtual water content of some selected primary food products is shown in Table 4.1. As the virtual water content of any product will vary depending upon the climate, the technology deployed and the yields achieved, the table provides the limits of the range reported for a number of selected countries by Hoekstra and Chapagain (2007) as well as the global average.

In the same paper, the authors report that the total global volume of water used for crop production at field level, i.e. before processing, amounts to 6390 billion m^3 per year. Rice constitutes the largest share,

Table 4.1 *Global average virtual water content of selected primary foods (m^3/tonne)*

Product	Range	Global average
Rice (broken)	1822–4600	3,419
Wheat	619–2421	1,334
Maize	408–1937	909
Soybean	1076–4124	1,789
Coffee (roasted)	5790–33475	20,682
Tea	3002–11110	9,205
Beef	11019–37762	15,497
Pork	2211–6947	4,856
Sheep meat	3571–16878	6,143
Chicken meat	2198–7736	3,918
Eggs	1389–7531	3,340
Milk	641–2382	990

Source: adapted from Hoekstra and Chapagain (2007).

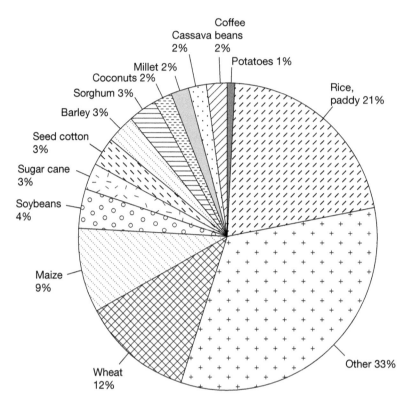

Figure 4.3 *Share of global field-level water use by different crops*
Source: adapted from Hoekstra and Chapagain (2007).

consuming 1359 billion m³ per year, or 21 per cent of total volume used for crop production, while wheat has a 12 per cent share. To put this in perspective, wheat occupies nearly 40 per cent more cropped area than rice and produces a little less in total volume of output (see Table 3.1). The breakdown of global water use by crops is shown in Figure 4.3.

It is clear from these data that, depending upon the composition of our diet, we eat a greater or lesser amount of virtual water. However, if many of the food items that we consume are imported from other countries, then we are also importing virtual water embodied in the production of those items. For water-scarce countries, the concept of virtual water has tremendous potential in guiding agricultural policy, for it would suggest the most rational course of action would be to import products requiring a lot of water in their production, rather than to produce them domestically. This would result in water savings, relieving the pressure on resources and allowing them to meet the needs of other users. Jordan, for example, now imports 60–90 per cent of its food, representing 5–7 billion m^3 per year of water, which is in contrast to the 1 billion m^3 per year it withdraws from domestic water sources (Roth and Warner 2008). On the other hand, for countries rich in water resources, the virtual water content of imported foods probably has little bearing on their demand for products that cannot be grown domestically as a consequence of climatic factors other than water availability. In this respect, the notion of virtual water can only go so far as a guide to domestic production and international trade.

Nevertheless, it is apparent that the international flow of virtual water within global trade is significant. Chapagain and Hoekstra (2008) undertook a comprehensive study of international virtual water flows for the period 1997–2001 and estimated this amounted to 1625 billion m^3 annually. Of this, 61 per cent related to international trade in crops and crop products, and 17 per cent to trade in livestock products. Industrial goods accounted for the remaining 22 per cent. Of particular interest from this study is the identification of the major exporting and importing countries of virtual water. Figure 4.4 provides summary data from a more comprehensive table developed by Chapagain and Hoekstra, which provides detailed accounts of virtual water imports and exports for twenty-four countries. Figure 4.4 lists the eight most significant exporters and importers of virtual water related to crop and livestock products only.

The data are revealing: Australia, a continent-sized country that comprises a great deal of desert and semi-arid environments, and which in recent years has been experiencing serious drought, is the world's largest *net* exporter of virtual water. The USA is, however, the largest gross exporter, that is, before subtracting its virtual water imports. India, too, emerges as a net exporter of virtual water, with 21.5 billion m^3 per year of water exported in crops, mostly to other countries in Asia. This is especially surprising given the state of India's groundwater resources: an article in the journal *Nature* highlights the unsustainable rates of extraction largely for irrigation in the country's north-western states of Punjab, Haryana and Rajasthan, where excess withdrawal of 109 km^3 occurred between 2002 and 2006 (Rodell *et al.* 2009). On a regional basis, Africa is a net exporter of water to other continents, particularly Europe. By contrast, the largest importer of virtual water – and by some

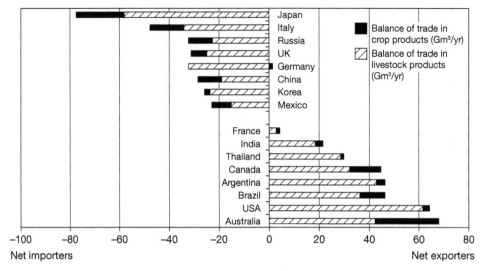

Figure 4.4 *Virtual water flows: leading exporters and importers*
Source: drawn from data in Chapagain and Hoekstra (2008).

distance – is Japan, which imports crop and livestock products containing over 77 billion m³ per year of water. On this basis, it could be asserted that Japan's resource and food security is achieved by drawing down water (as well as land) resources in other countries.

Water footprints

The concept of virtual water is useful when assessing the implications of international trade in primary foods: it helps reveal the hidden flows of water between countries and regions. However, if we wish to get a better understanding of an individual country's level of consumption of water, then the related concept of water footprint becomes especially valuable. In recent years the term "ecological footprint" has become widely used to represent the area of productive land and aquatic ecosystems required to produce the resources used, and to assimilate the wastes generated by, a population in a circumscribed area (country, region, city) (Wackernagel and Rees 1996). The water footprint, in contrast, indicates the volume of water required per capita to sustain a population. In order to calculate the total water footprint of a nation, it is necessary to quantify not only the water used to produce goods and services for domestic consumption (endogenous water), but also that virtual water which is embodied in imports of goods and services from other countries (exogenous water).

Perhaps unsurprisingly, the USA emerges with the largest water footprint per capita, 2480 m^3 per year, with the Mediterranean countries of Southern Europe not far behind. At the other end of the scale, the lowest water footprint is represented by China, with an average of 700 m^3 per year per capita. According to Hoekstra and Chapagain (2007), the size of the global water footprint is determined largely by the consumption of food and other agricultural products that comprise 86 per cent of the total, 73 per cent drawn from internal resources and 13 per cent from exogenous virtual supplies. The manufacture of industrial products, in contrast, accounts for just 9 per cent of the global water footprint, although this varies considerably between rich states (e.g. the consumption of industrial goods accounts for 32 per cent of the total water footprint of the USA) and developing countries (in India this sector accounts for just 2 per cent). Domestic water consumption for drinking and sanitation accounts for 5 per cent of the global water footprint.

Hoekstra and Chapagain (2007) identify four factors that explain high water footprints:

- The total volume of material consumption, which is related to gross national income; examples here include the USA, Switzerland and Italy.
- Climate – in areas of high evaporative demand, the water requirement per unit of crop production is relatively large, a situation facing Senegal, Mali, Sudan and Chad, amongst others.
- Water-inefficient agricultural practices are responsible for a low output per unit of water used; Thailand, Cambodia and Turkmenistan are amongst those guilty here.
- Finally, and significantly, a water-intensive pattern of consumption, especially where meat is a major part of the diet. The USA averages meat consumption of 120 kg per capita per year, a level that is three times the global average, but the countries of Western Europe are not so far behind. Given the rapid growth in meat consumption worldwide, the implications of this dietary change for water use is clearly of vital importance.

In drawing this section on water to a close, we can see that there is a long-standing and deep-seated relationship between societal development and the harnessing and management of freshwater resources. The capture and distribution of water for irrigation enabled higher levels of agricultural productivity, supporting greater numbers of people able to reach new cultural and technological achievements. As countries sought to extend their influence over distant lands, new foods were produced and transported around the world-shaping dietary habits. As demand grew, first in line with population numbers, then with rising material prosperity, food production required ever-greater quantities of freshwater for growing crops and grazing animals. An innate tendency toward the large scale meant bigger dams, larger pumps and greater rates of extraction. All too soon, we began to come up against some hydrological limits: the depletion of stocks (aquifers) and the unsustainable utilisation of flows.

Water footprint analysis and the related concept of virtual water help to sharpen our understanding of water-use efficiency in food production and to highlight potential problems in regions with limited water resources. Such analysis might encourage some countries to re-orient cropping systems away from products requiring a high endogenous water content to those requiring less, with the consequent shortfall in domestic demand for those products met by imports from countries with high positive water budgets and/or greater water productivity.

It also reveals significant inequalities between countries. For example, some African countries have hardly any external water footprint simply because they import so little; yet, despite experiencing water shortages themselves, they export crops with a high virtual water content, such as horticultural products, cocoa and cotton. Paradoxically, some European countries (e.g. UK, Netherlands) with abundant water resources of their own have a high external water footprint (accounting for up to 80 per cent of their total) given their imports of agricultural and industrial products. This situation reveals that an appreciation for the distribution of water resources worldwide has played no part in shaping the pattern of global trade, but ought to be factored into ongoing discussions that are otherwise driven by entirely financial considerations.

4. Peak oil and its implications

The current level of primary food production – as well as much of the post-farm-gate processing, manufacturing, distribution and retailing of food, as chapter five shows – owes a very great deal to the use of fossil fuels. Chemical fertilisers are manufactured from natural gas; oil drives field machinery, irrigation pumps and post-harvest equipment. These hydrocarbons derive from the ability of algae and plants to capture solar energy many millions of years ago, to bloom and then to die. The decay of this organic matter, buried deep under layers of sediment then compacted and heated over a prolonged period, became natural gas or crude oil depending upon the temperature and pressure to which it was subjected.

Were fossil fuels to be in infinite supply, there would no reason to suppose that the global agri-food system in its present state would not continue to rest upon an abundance of cheap energy. Unfortunately, there is a growing realisation that the evidence demonstrates otherwise: that we are at, or close to reaching, a situation of "peak oil", where we have used up approximately half the geological endowment of conventional oil and natural gas. This would suggest a near future of much tighter energy markets as supply is unable to meet demand and prices will rise. This will have direct and fairly immediate repercussions for primary food production, as events during 2008 demonstrated (discussed in chapter six).

Some of the most salient facts are as follows:

- The biggest oilfields in the world were discovered more than half a century ago on the Arabian peninsula.
- The peak of oil discovery was 1965, and apart from the discoveries of the Prudhoe Bay field in Alaska and the North Sea during the 1970s, there has been a relentless fall in discoveries since that time.
- An estimated 80 per cent of all oil used in the world comes from fields discovered before 1973.
- Since 1981, we have been using more oil than has been found, and the gap between discovery and consumption has been growing wider with every passing year. In 2007, the world consumed six barrels for every one that was discovered.
- The world is consuming 84 million barrels of oil each day, or 30 billion barrels per year. (A barrel comprises 42 US gallons, 35 UK gallons or 159 litres). Even a passing acquaintance with geology will suggest that fossil fuels cannot last indefinitely (Campbell 1997, 2003; Leggett 2005).

At one level, peak oil reflects the circumstances to which all individual oil fields generally subscribe. That is, when initially tapped and under its own pressure, oil flows freely on a rising curve of production until reaching a maximum point, to be followed, as pressure falls, by a declining return until exhaustion. Plotting such output on a graph provides a symmetrical, bell-shaped curve. Aggregating the individual bell curves for each oilfield in the lower forty-eight states of the USA was a task undertaken by geologist M. King Hubbert in 1956. Plotting the production rates over the preceding 100 years, and using a generous estimate for ultimate recoverable oil, Hubbert calculated that US production would peak in 1971. Although widely ridiculed within the industry, peak oil arrived in the lower forty-eight states a year earlier than Hubbert's prediction (Leggett 2005). In the United States, with its largely unconstrained market driven by the principles of free enterprise, the pattern of oil production was remarkably smooth and symmetrical. Elsewhere, however, political interventions in the oil market – such as that exercised by the Organisation of Petroleum Exporting Countries (OPEC) in 1973 when they restricted supplies in order to effect a substantial increase in crude oil prices – would result in more uneven curves; wars, technological change and other unanticipated events also have an effect on production patterns. In the UK, North Sea oil production peaked in 1999 and output is now around 43 per cent lower; Norway's production, too, is now 25 per cent less than in 2001; while the Prudhoe Bay field in Alaska produces just a third of the oil that it did two decades ago (Mouawad 2008).

Future projections

The key issue, however, is: just how much oil is there? We will put natural gas to one side for the moment and concentrate on what is called "conventional" oil. This is the lightest crude that has dominated production to date, and has been found in the largest reservoirs that were easiest to locate and to exploit: close to the surface, often on dry land, and flowing under their own pressure. Given that geologists, both on the ground and remotely using the latest satellite technology, have mapped the world extensively, one might expect a consensus on the total ultimate recoverable resource. The world of oil, however, is one of Machiavellian politics, where producer countries and oil companies revise estimates of assets for purposes other than scientific transparency. Fortunately, a network of independent analysts, led by a retired petroleum geologist, Dr Colin Campbell, have worked assiduously over the past fifteen or so years to strip away the obfuscation and political spin associated with this data. According to the Association for the Study of Peak Oil & Gas (ASPO), founded by Dr Campbell, the best estimate for the total endowment of conventional oil prior to its exploitation is a little less than 2 trillion barrels (2×10^{12}) (Campbell and Laherrère 1998). Of this, we have used to date around 950 billion barrels, which leaves a little over 1 trillion barrels to exploit. While this may seem sufficient to keep us powered for many years ahead, the geological reality casts something of a shadow:

- Much of the remaining oil is heavier and of lower quality, with higher sulfur content, and requires more effort to refine.
- It is generally found in smaller deposits below the category of giant fields, defined as containing more than 500 million barrels or capable of pumping 100,000 barrels per day. The 500 giant fields comprise just 1 per cent of the total number of oil fields in the world, yet contribute more than 60 per cent of global output (Höök et al. 2009). Smaller fields are more costly to exploit given their size.
- The new fields are likely to be found offshore, and increasingly in deep water, where the difficulties and costs of extraction are very much greater.
- The remaining 1 trillion barrels also includes the residual oil in the original fields. This is the oil that no longer flows under natural pressure, but must be pumped out using water or gas or by fracturing the rock strata. Even with such advanced techniques as directional drilling, the most oil that can be extracted from a reservoir is 60 per cent of its total volume, and for many fields it is a good deal less (35–40 per cent). The Ghawar field in Saudi Arabia, the largest oil deposit ever discovered, with 87.5 billion barrels, still accounts for 5 per cent of the world's total daily production, more than sixty years after its discovery. Yet to keep the oil flowing, 7 million barrels of seawater are injected into the reservoir every day (Campbell 2003; Leggett 2005).

Figures 4.5 and 4.6 set out the evidence and demonstrate why it is possible to assert that we are at peak production. Figure 4.5 shows a bar chart of past and possible future discovery and plots a production line to the present: its future trajectory will probably be in a downward direction, but the rate of its decline will depend upon a host of factors. For example, greater efforts now to maximise recovery of oil in order to maintain total output will see much sharper declines in the future. Figure 4.5 demonstrates very clearly the points noted above – since the early 1980s we have been using more oil than has been found, and with each passing year this disparity between consumption and discovery has been growing.

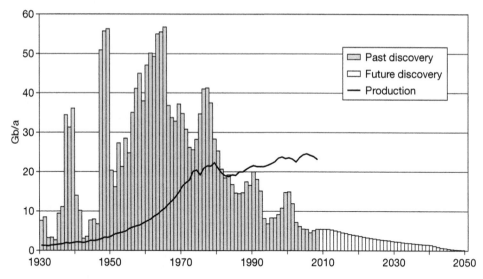

Figure 4.5 *Discovery and production of regular oil*
Source: Colin Campbell, personal communication.

Figure 4.6 sets out the depletion curves as derived by Dr Colin Campbell from his analysis of industry data, and include conventional and non-conventional oil as well as natural gas. Regular (conventional) oil has dominated output to date and will continue to do so into the future, yet the evidence points to a likely peak in production around 2005. By comparison, the likely contribution of non-conventional sources by volume appears to be relatively modest. This is especially salient when set against the enormous financial and technological demands required in their exploitation (e.g. drilling in deep ocean sea beds or in the Arctic), as well as the potentially catastrophic environmental impacts likely to result. That we have reached peak production beyond which any further increase in output of regular oil is now widely regarded as extremely unlikely; but with rising levels of demand, especially from China and India, oil prices are set to move generally upward. This is why there has been such enthusiasm for non-conventional sources of oil, such as the Canadian tar sands ("heavy oil"), as well as the development of biofuels. Box 4.5 sets out some of the issues related to the exploitation of tar sands. The issue of biofuels has even greater implications for primary food production, and is explored in a little more detail below. A final observation with regard to Figure 4.6 is drawn from Colin Campbell, who reminds us not to be mesmerised by the date of peak. What matters, he says, is to recognise

Box 4.5

Canada's tar sands: dirty oil

An estimated 2 trillion barrels of oil exists in the form of tar sands – about the same quantity as conventional oil – although considerably more effort is needed to retrieve it. Much of the current production is taking place in the Province of Alberta in Canada, creating what has been called a modern-day gold rush. But there are no individual prospectors here, only consortia of major energy companies with a stake in the world's largest energy project.

The constituent of tar sands is bitumen, a dirty, heavy oil. There are two principal methods of extraction: open-cast mining, which involves the removal of up to 75 metres of overburden (soil, clay and sand) to reveal the seam of oil sands around 50 metres in depth; or *in situ* extraction, which accounts for 80 per cent of production, involving high-pressure injection of steam or solvents such as naptha. Fire may also be used to heat the bitumen and coax it to the surface.

Processing involves washing the material in hot water in order to separate the bitumen from the sand; the bitumen is then processed further to make crude oil. The polluted water used in processing – three barrels for every barrel of bitumen – is then pumped into enormous tailing ponds, which cover over 100 km^2.

The environmental balance sheet of this industry does appear to be truly daunting. The volumes of freshwater being extracted for processing are causing serious concerns for the integrity of the hydrological resources of the region. The tailing ponds are leaving a terrible environmental legacy, with huge volumes of highly toxic wastewater leaching into groundwater and the river systems. Propane cannons are used to deter migratory flocks of birds from settling on the ponds, mistaking them for freshwater feeding grounds, leading to almost immediate death. Reports on human health impacts in the wider region, particularly on First Nation peoples who continue to hunt and fish for food, appear deeply worrying. This development has become Canada's fastest-growing source of greenhouse gas emissions, with oil from tar sands generating 20–30 per cent more greenhouse gas emissions than from conventional oil. Meanwhile, 15–20 per cent of Canada's annual production of natural gas is used to provide the process energy. This has been described as reverse alchemy: turning clean-burning natural gas into dirty, carbon-heavy oil.

Sources: adapted from Leggett (2005); Environmental Integrity Project (2008); Government of Alberta (2010); Polaris Institute (2010).

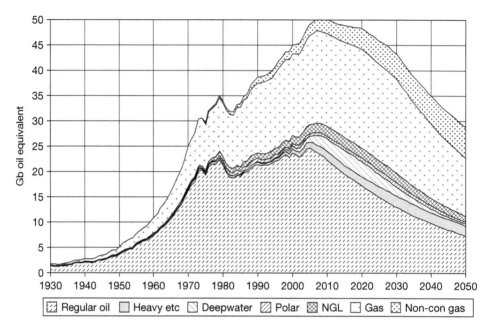

Figure 4.6 *The depletion curve for oil and gas*
Source: Colin Campbell, personal communication.

the relentless decline that follows the peak, representing a future of ever-falling output of regular oil (and gas), which unconventional sources will only partly and temporarily (and at significant cost) help to buffer.

Biofuels

In contrast to the development of non-conventional oil as a way of extending the supply of petroleum, biofuels have been heralded as a means to offset dependence upon petroleum. Both in Europe and in the United States, there has been considerable interest in developing energy sources from biomass, that is, plant material derived from photosynthesis and processed to produce a gas, liquid or solid that can be burnt to release useful energy, for example in an internal combustion engine. While biofuel technologies pre-date the Second World War, the low price, ubiquity and high energy density of oil products made them uncompetitive – at least until the oil-price rises and supply restrictions of the 1970s, when some countries began seriously to explore ways of reducing their dependence on imported petroleum. One such country was Brazil, which built much of its early economic development around the automobile, and which quickly began to develop an ethanol industry derived from sugar cane that, by 2006, produced over 16 billion litres, around 20 per cent of the country's total fuel consumption (Shurtleff and Burnett 2008). Today, all automobiles in Brazil are required to burn an E25 blend of gasoline (75 per cent) and ethanol (25 per cent), and there is a growing fleet of "flex-cars" that run on any mix up to E100 pure ethanol fuel. The successful transition in implementing such a measure and the enormous foreign exchange savings made have encouraged other states to follow suit, and world ethanol production has grown accordingly (Figure 4.7).

The United States has embarked upon a rapid expansion of its own ethanol programme, building new refineries to process maize (corn) into fuel. By 2008, 134 ethanol plants were in production compared with 68 in 2003. In 2007, some 81 million tonnes (or 20 per cent of the US maize harvest) was diverted to ethanol production, yielding

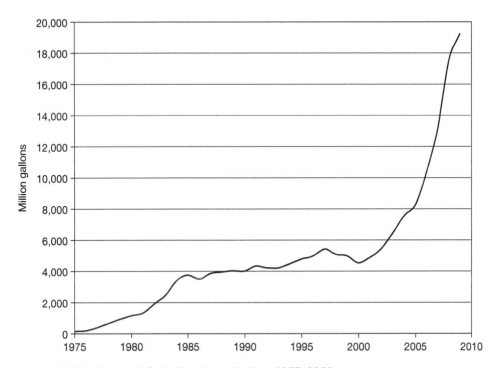

Figure 4.7 *World annual fuel ethanol production, 1975–2009*
Source: Earth Policy Institute, www.earth-policy.org.

6.5 billion US gallons (around 24 billion litres). The industry is well ahead of federal targets – for example, the Energy Independence and Security Act of 2007 requires 8 billion gallons of renewable fuels to be blended into the country's fuel supply in 2008 – and was set to achieve an installed capacity of over 13 billion gallons in 2008. As oil prices rise, the economics of ethanol manufacture become increasingly attractive, made more so by government subsidies paid to distillers.

In Brazil, ethanol is competitive with gasoline/petrol when oil prices are at $35–40 per barrel. Brazil also has a major comparative advantage insofar as sugar cane produces twice as much ethanol per hectare than corn (up to 8000 litres/ha compared with 4000 for corn). Critically, the energy balance – that is, the ratio of the energy obtained from ethanol to the energy expended in its production (energy returned on energy invested, EROEI) – is far superior in the case of sugar cane: up to ten times, compared with up to 1.6 times for corn. Besides many other advantages, perhaps the most critical factor of concern here is that sugar cane is not a staple food and, although occupying arable land, is not otherwise distorting market prices. Corn, on the other hand, is a major food ingredient within the United States, and corn exports provide consumers in distant markets with subsistence. As ethanol distilleries agree contracts for their feedstock with farmers, they are competing directly with those who would otherwise eat that corn (an issue to which we return in chapter six).

In the case of the European Union, where production of ethanol is modest compared with the USA and Brazil, there has been more interest in developing biodiesel fuels using vegetable oils as feedstock. A variety of agricultural crops have been harnessed, including those widely grown in Europe (rapeseed, sunflower) and those imported from the tropics (oil palm, coconut, peanut). Domestically grown soybean accounts for 80 per cent of US biodiesel production. Because of strong environmental criticisms that have drawn attention to tropical deforestation in countries that have sought to expand production of commodities such as oil palm for European biodiesel, the EU has reconsidered its initial target of achieving a 5.75 per cent share of all transport fuels with biofuels by 2010 and 10 per cent by 2020. Instead, it has set new targets encouraging the development of waste, non-food crops and what are referred to as second- and third-generation biofuels. Second-generation biofuels focus upon non-food biomass crops, such as quick-growing cellulosic grasses and woody shrubs (e.g. miscanthus and switchgrass) to produce ethanol, and jatropha for biodiesel. Third-generation biofuels, which are clearly at an even earlier stage of research and development, involve the use of algae. Future editions of this book may need to devote space to algaculture as a source of food and energy.

Initially, there was some enthusiasm for agro-biofuels, which were regarded as environmentally benign, offering opportunities to reduce dependence upon foreign oil and thereby enhance energy security, and even to improve farm incomes. More recent analysis, however, has been guarded, if not critical, particularly with regard to claims that agro-biofuels are at least carbon neutral. As discussed earlier in this chapter, the conversion of natural vegetation cover to biofuel production is likely to liberate larger amounts of CO_2 locked up in that vegetation than can be captured by the new crop, and the process may significantly reduce the carbon-sequestration capacity of soils. Furthermore, far from creating new opportunities for farm incomes, large-scale biofuel cultivation is likely to further the elimination of small farmers in the interests of efficiencies of scale.

Consequently, the challenge of peak oil requires an urgent evaluation of our predicament, laced by an understanding that whatever alternative is developed is unlikely to possess the ubiquity, flexibility and price of oil. Such alternatives need to meet more stringent criteria than those that seem to exist with regard to the development of tar sands: new energy sources should not generate such enormous environmental impacts, nor should they compromise global food production capacity. The challenge of peak oil ought to present an opportunity to look more closely at our assumptions in continuing the extravagant patterns of energy consumption that we have enjoyed hitherto. In the realm of food, it might give us pause to consider the degree of technological and institutional lock-in to oil-based production, distribution, retailing and consumption, and offer an opportunity to think more closely about our dietary options. The consumption of livestock products would necessarily be high on this list.

5. Livestock and the rising demand for meat

Livestock's impact on the environment is already huge, and it is growing and rapidly changing. Global demand for meat, milk and eggs is fast increasing, driven by rising incomes, growing populations and urbanization.

(Steinfeld et al. 2006: 3)

Meat consumption

We are naturally omnivores, physiologically evolved and culturally adapted to eating a wide range of foods, including those of animal origin. It has been argued that animal foods – whether scavenged, collected, hunted or produced from domesticated species – have been a universal feature of human behaviour, with valuable nutritional and social implications. An issue that is of growing contemporary concern, however, is whether we are consuming animal products, particularly meat, at levels that are harmful to human health and to the environment. Livestock products provide a range of nutrients, including protein, minerals such as iron and calcium, vitamins including B12, and fat. However, nutritional research over the past two decades has generated a body of scientific evidence that has demonstrated the adverse effects of animal fats (related to cardiovascular disease) and animal protein (related to cancer). The environmental consequences of intensive livestock-production systems are also significant and wide-ranging, with meat production, for example, estimated to account for 18 per cent of global warming, an effect greater than the transport sector worldwide.

Pimentel and Pimentel (2003) have asserted that an estimated 2 billion people worldwide live on a meat-based diet, while an estimated 4 billion people live primarily on a plant-based diet. Yet the past decade has witnessed a major expansion of meat eating such that, today, the Pimentels' ratio might well be in balance. This is a consequence of rising levels of disposable income as well as dietary change that arises from people moving to the cities. It has long been recognised that, at low levels of income, indigenous staples such as cereals and tubers dominate food intake. Beyond a certain income threshold, however, there is a marked increase in meat consumption as rising disposable income allows for greater dietary diversification and people opt for higher sources of protein. With a marked upturn in economic fortunes in China, India, some other Asian and Latin American countries, meat consumption has grown rapidly. This "nutrition transition", as it has been termed, occurring within a single generation, has seen shifts in some parts of the world from situations of widespread undernutrition to aspects of overnutrition. This is an issue to which we return in chapter six.

Table 4.2 provides a summary of meat consumption per capita by region. It shows the sharp increase in meat consumption amongst the middle-income group of countries, which are those experiencing the steepest improvements in household income. Here, meat intake has grown from around 16 kg per person in 1970 to 46 kg per person in 2002. This is in sharp contrast to the low-income countries, which as a group have registered a barely noticeable improvement, with a rise from 6.7 to 8.8 kg between 1970 and 2002. One country that has moved between these economic categories during this period, and that can demonstrate a level of meat consumption to match, is China. Although Keyzer *et al.* (2005) note some discrepancies in official Chinese statistical reporting of consumption levels, it is clear that the trend in meat eating is sharply upward, rising from around 9 kg per person in 1970 to a reputed 52 kg in 2002, with pork comprising the most significant element. If the FAO data are approximately correct, not only did meat consumption in China double in little more than a decade (from 26 kg per capita in 1990), but by 2005, China consumed more meat than the entire population of the world in 1961 (Weiss 2007). Amongst the high-income countries, on the other hand, meat consumption has risen more slowly but steadily, although there is some considerable variation between countries, even within Europe. While the UK recorded a level of per capita meat intake of around 80 kg in 2002, in Denmark the figure was 146 kg – exceeding even the USA's consumption of 125 kg.

Table 4.2 *Meat consumption per capita by region (kg per person per year)*

Region/class	2002	1990	1980
Asia	28	17	11
Europe	74	n/a	n/a
North America	123	111	107
South America	70	50	47
Sub-Saharan Africa	13	14	15
High-income countries	94	85	79
Low-income countries	9	8	7
Middle-income countries	46	29	22

Source: World Resources Institute, http://earthtrends.wri.org/searchable_db (figures rounded by the author; n/a, not available).

Livestock systems

Large-scale and intensive livestock production really began to take off from the 1950s. Prior to this, the raising of animals was much more closely tied to the availability of local resources and a degree of multifunctionality at farm level, such as the grazing of crop residues in fields as part of rotations in order to avail of the manure, the use of pigs to root and break up the sod after a long fallow, or the conversion of organic wastes to meat and eggs. The herding of ruminants (cows, sheep, goats) also involved local daily or more long-distance seasonal movements in order to take advantage of different grazing resources; for example, patterns of transhumance in mountainous regions with cattle and sheep taken to the high summer pastures (where their milk was turned into butter or cheese), or long-distance pastoralism, a particular feature of nomadic cultures, which proved an effective evolutionary adaptation to the sparse grazing of arid environments. In all these cases, livestock products were central to human survival, with milk and its fermented derivatives, blood and meat contributing variously to the human diet.

Although grazing land takes up over one-quarter of the ice-free terrestrial surface of the planet, livestock also account for around one-third of arable land that is given over to the production of animal feeds. The expansion of pasture in regions where there was no previous history of livestock grazing has seen considerable environmental destruction, perhaps the most controversial being the level of deforestation in Amazonia and Central America. During the 1970s, the World Bank substantially increased its loans to Central American countries for agricultural and rural development, ostensibly to improve basic human needs and poverty alleviation. But the majority of loans went to support the production of beef for export (largely for US hamburgers), creating precious few jobs and limited livelihood opportunities for the rural poor. In Brazil, the push into Amazonia – though now joined by the expansion of soybean – was initially driven by ranchers using cattle to displace small-scale farmers and to control, as cheaply as possible, large areas of land. Consequently, although it is easy to represent cattle as primary instruments of environmental destruction, they are really ciphers of more powerful economic and political drivers of change. Box 4.6 reminds us that cattle culture has deep historical roots, if a poor modern-day reputation.

According to FAO figures, there are an estimated 1.3 billion cattle in the world, with the greatest numbers in India, Brazil and across Sub-Saharan Africa. A full breakdown of livestock numbers by world region is provided in Table 4.3. If the South accounts for three-quarters of the global cattle herd, it has a lesser share in the total number of pigs and poultry, which are increasingly produced in intensive operations, and a slightly higher share in small ruminants (sheep, goats, camelids), which are most commonly grazed more extensively. The bald figures can disguise enormous differences in systems of production and the economies they support.

For the cattle-herding peoples of East, Southern and Sahelian Africa, as well as for pastoralists elsewhere, grazing ruminant animals across dry and challenging environments is an effective means of survival, provided the herds and flocks move regularly between pasturelands and waterholes. The use of established migratory routes ensures that grazing

---

Box 4.6

Bovine culture

In an impassioned book, Jeremy Rifkin traces the deep-seated significance of the bull and the cow in human culture: from its first domestication in Mesopotamia and its yoking to a plough, the bull became a hugely important symbol of divinity throughout the Mediterranean (consider, for example, the Minoan civilisation). Bovine culture then spread, via the Nile Valley, from Egypt deep into Africa, creating pastoralist societies from the east to the very south of the continent. Meanwhile, the eastern migration of Eurasian nomads brought the cow to India, where it became, over 2000 years, a sacred animal, a goddess of abundance. This is in sharp contrast to Spain, where it is the machismo of the bull that is celebrated, at least by part of the population, through the sacrificial rituals of the bullring. The conquest of the Americas also brought its "cattlising", with both conquistadors and catholic priests introducing the animals to new lands from the Pampas of Argentina and Uruguay through large swathes of Venezuela and Mexico to Texas and Florida. The extermination of the buffalo across the Great Plains of the USA by 1870, as well as the First Peoples who lived alongside them, opened up the range to cattle ranching. With the railway providing their transport to the stockyards of Chicago, and with the development of refrigerated shipping by the 1880s, American beef found its way onto European tables.

Cattle have consequently played a critical role in changing the ecology of many habitats around the world, especially in the Americas. It is little wonder that, for Rifkin, the creation of the modern cattle complex is a truly malevolent force, responsible for a major desecration of the world's ecosystems.

Source: adapted from Rifkin (1992).

Table 4.3 *Population of the principal commercial livestock categories by world region*

Region	Population (million head)			
	Cattle	Pigs	Small ruminants	Poultry
North America	111	73	9	2,059
Latin America and Caribbean	358	77	116	2,256
Brazil	177	32	24	878
Europe	101	165	142	1,329
Commonwealth of Independent States	58	31	60	558
West Asia and North Africa	32	0.7	227	1,263
Sub-Saharan Africa	213	20	370	862
South Asia	246	15	299	701
India	191	14	181	377
East and South-East Asia	153	529	346	5,995
China	104	452	289	3,830
Oceania incl. Australia	65	8	266	199
Total: North	327	285	400	4,519
Total: South	984	632	1,322	10,628
Total: world	**1,311**	**918**	**1,722**	**15,147**

Source: assembled from data in Steinfeld *et al.* (2006).

resources are used in a sustainable way, allowing for their full recovery between visits. With the end of colonial rule came the creation of states demarcated by borders drawn across traditional migration routes. The imposition of head taxes, the introduction of the monetary economy that led to the breakdown of traditional bartering arrangements, and efforts to sedentarise nomadic pastoralists all contributed to the creation of an emerging ecological problem, which was made worse by the drilling of wells to create permanent waterholes. As herds grew in these settled areas, overgrazing has led to significant problems of desertification. In the meantime, pastoralists, for whom their animals represent their sole assets as well as their cultural identity, encroach onto the lands of farmers in their search for grazing, contributing to escalating tensions and conflict in a context of resource constraints.

Factory farming

If the arid regions of Sub-Saharan Africa, where thin cattle browse on thorny shrubs, illustrate the timeless relationship between animals and the availability of local resources, the past half-century has seen the severing of this link in the development of modern livestock production. With the emergence of grain feeding of animals, beginning in the USA in the 1950s and spreading quickly to Europe, the Soviet Union and Japan in the 1960s, suddenly there were no local resource constraints to the number of animals that could be raised. Although this has had an utterly transformative effect on the production of monogastrics (pigs and poultry), it has also had a large impact on the cattle sector in the USA. In contrast to the scarcity of grazing in Sub-Saharan Africa, the overcrowded feeding lots of Kansas and Colorado, where cattle are fattened on a diet of corn and supplements, makes us appreciate the immensely adaptable nature of the bovine. Box 4.7 draws from Michael Pollan's account of the feedlot system, and describes the way in which up to half of all beef is produced in the United States today.

But the rise of feedlot production for beef has been outstripped by developments that have transformed the "farming" of pigs and poultry. High stocking densities, confinement practices and measures aimed at "speeding up" the growth cycle of animals have become standard procedures for those production units officially known in the United States as CAFOs (confined, or concentrated, animal feeding operations), elsewhere colloquially as "factory farms". In recent years, organisations concerned with animal welfare (e.g. Compassion in World Farming) have worked hard to bring to wider public attention the conditions under which poultry (layers, broilers) and animals (pigs, veal calves) are kept. But the intensification of livestock production, driven by the imperative of reducing costs and thereby making meat and dairy ever cheaper, has brought little sustained public opprobrium. Most of the criticism at such production practices occurs only episodically at times of outbreak of a zoonotic disease (one that is capable of being passed from animals to humans).

Modern animal production did not set out to be cruel or wasteful, but to be efficient, to get more output for less input. Above all, modern methods were designed to reduce labour, with other inputs either freely available (air, water) or relatively cheap (energy, feeds) (Gussow 1994). As part of the process of intensification and specialisation described in chapter two, farms began to lose the small-scale and multifunctional aspects of animal husbandry in which a few pigs and small flock of hens were an almost ubiquitous feature of the average European or North American farmyard. Gradually, a combination of market incentives and public policy regulation (on the grounds, ironically, of animal health and food safety) encouraged specialised and large-scale production in

Box 4.7

Feedlot beef

Born on a ranch on the Great Plains, a calf spends its first six months eating grass, which is what it is physiologically evolved to do. The animal's rumen, the first of four compartments in a bovine stomach, contains powerful bacteria capable of breaking down and digesting the cellulosic material that is utterly indigestible to humans. Growing meat on grass makes ecological sense, particularly in places like South Dakota, where the land is incapable of growing row crops without large amounts of irrigation and chemicals. Cattle that feed on pasture, or on crop residues, or by-products of food processing are able to metabolise such material and therefore do not compete with humans directly for food. However, calves born on the grasslands are introduced to maize (corn) at six or seven months, and by nine or ten months are part of the huge bovine population that comprises the American feedlot beef production system.

Consequently, feedlot cattle – or at least the specific steer which Pollan bought on a ranch and followed through the system – are given a mixed diet of around 15 kg/day, consisting of corn, liquefied fat (possibly rendered from animal carcasses), protein supplement (comprising molasses and urea), silage and antibiotics. The latter are critical in keeping the animal healthy – or, at least, healthy enough until its slaughter after five months – as the diet puts enormous strain on the animal's capacity to digest a diet it was never designed to consume. Yet, while it converts this daily intake into 1.5 kg of body weight, the greater part of this intake is converted into a waste stream, some of which is drained off into lagoons, some of which simply finds its way down into the groundwater.

Pollan makes clear the connectedness of the feedlot: back to the fields that provide the bulk of the cattle feed, to the vast monoculture of corn that is fed by "a steady rain of pesticide and fertiliser"; and, following this further upstream, to the oil fields of the Persian gulf. Downstream of the cattle feedlot is, metaphorically and physically, the Mississippi Delta and the Gulf of Mexico, where 5800 square miles (according to NASA) of coastal waters have been rendered lifeless because of the excessive nutrient loading that derives from fields and feedlots.

Pollan's calculation was that his purchased steer consumed nearly a barrel of oil during its life of 15 months, enabling it to grow from a 36 kg calf to a beast of 545 kg. Though inefficient, bovine metabolism has been transformed into another dimension of food industrialisation driven by fossil fuels.

Sources: adapted from Pollan (2006); NASA (2009).

dedicated units. Such production systems lent themselves to new opportunities for agro-industrial appropriationism, supplying dedicated feeds, medicines, housing, waste-management equipment, and scientific expertise in producing higher-yielding and faster-maturing breeds. Little wonder that some of the world's largest grain companies (Cargill, ConAgra) began to develop strong commercial interests in livestock production. Recalling the quotation from Page (1997) in chapter two regarding hog production in the USA, the diffusion of capital-intensive techniques resulted in a dramatic reduction in the number of farms raising pigs, and a sharp increase in the average number of pigs per farm. Almost everywhere that dedicated pig units are created, the numbers of animals are no longer counted in their tens or dozens, but in their hundreds or thousands. Consequently, factory farms have come to account for around 40 per cent of global meat production by volume, with around three-quarters of the world's poultry, 68 per cent of egg production, and about half of the world's pig meat produced in confined feeding operations (Weiss 2007).

As the concentration and stocking density of factory-farmed animals has increased, one can reflect that these creatures seem to have lost their sentient character to become much like machines housed in rural factories. Animal welfare organisations have drawn attention to confinement practices that include the use of cages for battery hens and pens for nursing sows that do not even allow sufficient room to turn; and have highlighted common surgical procedures including debeaking, docking tails, clipping piglets' teeth, and castration without anaesthetic. Managing reproduction in pigs can involve surgically relocating a boar's penis so that it can identify sows in a fertile state but allows the semen to be collected by a handler, which can then be carefully allocated in order to artificially inseminate twenty sows. Above all, the extraordinary effort to drive down costs of production in order that meat can become an ever-cheaper food staple has led to a high-throughput, high-speed system. All of this in order that pork (and chicken, for parallel practices occur in the poultry industry) is available at ever-cheaper prices. Box 4.8 provides some further details of intensive pig production in the USA.

Environmental impacts

Given the global spread of intensive livestock production systems capable of producing the high volumes of meat demanded by societies experiencing improved economic prosperity, it is clear that animal

---

Box 4.8

A pig's life

More than 90 per cent of pigs raised for meat today are raised indoors in crowded pens of concrete and steel. They never get to go outside or root around in pasture, and don't even have straw to bed down in. The most tightly confined of all are the breeding sows. Under the factory's rigid production schedule, they are made to produce litter after litter as quickly as possible, which means that they are pregnant for most of their lives. During their pregnancies, which last about 16 weeks, most American sows are confined in "gestation crates" – steel-barred crates or stalls just a foot longer than their bodies, and so narrow that the sows cannot even turn around. Of the 1.8 million sows used for breeding by America's ten biggest pig producers, about 90 per cent are kept in this manner.

When the time comes to give birth, they are also confined in a farrowing crate. [This] keeps the sow in position, with her teats always exposed to her piglets. She is unable to roll over – and this, the defenders of the crate say, ensures that she will not roll on top of, and perhaps smother, her piglets.

In Europe, widespread public concern about the close confinement of sows led to the European Union asking its scientific veterinary advisory committee to investigate the impact of gestation crates on the welfare of sows. Pigs like to forage and explore their environment. In a stall, they cannot, and after a time they become inactive and unresponsive, or carry out meaningless, repetitive motions that are signs of stress. The scientific committee concluded, "sows should preferably be kept in groups". The EU is now phasing out sow crates by the end of 2012, and requiring that sows be given straw that they can play around with, to reduce the stress of boredom. Even before the new law comes into effect, Britain and Sweden acted to ban sow stalls. All of the 600,000 breeding sows in Britain now have, at least, room to turn around and can interact with other pigs.

Source: adapted from Singer and Mason (2006), 45–7.

numbers have long outstripped the capacity to feed them on locally available crop residues, by-products and food waste. Instead, increasing volumes of cereals and oil seeds are produced for dedicated use in the animal feeds sector, and there are growing concerns that this represents a source of considerable uncertainty for long-term human food security. In 2002, the livestock industry accounted for 670 mt of grains, over 200 mt of processed oil seeds and pulses, and 150 mt of roots and tubers (Steinfeld *et al.* 2006). The particular combination of feed concentrates depends naturally upon the livestock, but also upon the comparative cost advantages within different regions. For example, maize is the dominant cereal used in animal feeds in the USA, Brazil and China, while wheat and barley prevail in Canada and Europe. According to Pimentel and Pimentel (2003), US livestock consume more than seven times as much grain as is consumed directly by the entire American population; the amount of grains fed to US livestock is sufficient to feed about 840 million people following a plant-based diet. Grain fed to livestock as a proportion of the total grain consumed now accounts for some 60 per cent of the total grain consumption of the USA, 69 per cent in Australia, and 73 per cent in Canada.

Soybean meal is a ubiquitous element of pig and poultry rations globally, and demand for it over the past four decades has increased at a faster rate than total meat production, suggesting a net increase in its use per unit of meat. It was noted above how soybean was strongly associated with deforestation in Amazonia and how quickly output has increased, with Brazil, Argentina and the USA today accounting for about 80 per cent of global production. This places importing countries at something of a disadvantage when it comes to specifying standards. For example, the UK imports about 90 per cent of its 3 mt of soymeal from Brazil and Argentina. However, these two countries are well advanced in converting their soybean production to genetically modified (GM) varieties, while the UK, as part of EU policy, seeks to import only non-GM soy. The processing of soybeans, it should be said, gives rise to about 19 per cent oil, which can be used elsewhere in the food industry, and 74 per cent meal, which is used for animal feed.

While it has been argued that feeding grains to livestock represents a buffer through which grain surpluses can be utilised, thus supporting farm-gate prices, the more convincing argument points to the spread and intensification of cereal and oil seed farming to produce low-cost inputs for the feedstock industry. The bottom line is that the conversion of plant to animal protein is inefficient and involves a significant net loss. According to Steinfeld et al. (2006), livestock consume 77 million tonnes (mt) of protein in feedstuffs that could potentially be used for human nutrition, whereas the products that livestock supply contain only 58 mt of protein. In dietary energy (calorie) terms, the net loss is even higher, particularly when calculations include all the fossil energy inputs into the cultivation of feed crops. According to Pimentel and Pimentel (2008), chicken broiler production is the most efficient, with an input of 4 kcal of fossil energy per 1 kcal of protein produced. Thereafter, input–output ratios for other livestock products produced from concentrates deteriorate quickly, with pork at 14:1, eggs at 39:1 and beef at 40:1. As they argue, if animals were fed on good-quality forage rather than grains, the energy inputs could be reduced by half.

In a context of rising global demand for meat, there are consequently serious anxieties about how the increased demand for feedstuffs – variously estimated at between 1 and 1.9 billion t over 1999 levels by 2030 – will be met while also providing for an extra 2 billion more people. There is also concern about the various waste streams derived from intensive livestock operations. With livestock estimated to contribute around 18 per cent of the global warming effect – principally derived from a 37 per cent share of methane emissions and 65 per cent of nitrous oxide – the link to climate change is well established. Regionally and locally, livestock wastes are a major source of ammonia and create considerable damage to air quality, seriously affecting the health and wellbeing of residents in proximity to feeding operations and slaughter facilities. Water pollution, too, is strongly associated with the inadequate management of animal manure, of which an estimated 500 mt was produced by feeding operations in 2003. As Michael Pollan reminds us, one of the striking things that any intensive animal-feeding operation does is to take the elegant ecological solution of livestock manure applied to fields for fertilisation and neatly divide it into two new problems: a fertility problem on the farm (which is resolved with chemical fertilisers), and a pollution problem arising from the livestock.

It is the scale of this pollution problem that has become a serious public health concern. The reasons why this waste is no longer valuable nutrient material that can be returned to the soil to perform a fertilisation function are as follows:

- Because of the high level of antibiotics routinely used as prophylactics to keep animals from falling ill, there are high levels of pharmaceutical residues in the waste stream.
- This has also led to the development of antibiotic-resistant pathogens that can prove very dangerous to human health, as normal broad-spectrum antibiotics (e.g. tetracyclines) are no longer effective.
- The widespread use of growth hormones to speed weight gain in animals also creates a residue problem in the wider environment and, together with certain forms of pesticides, has created hormonal changes and mutations in aquatic life such as frogs and fish.

- The waste stream can contain zoonotic pathogens (parasites, bacteria, viruses) that can be harmful to human health.
- The waste stream comprises effluents high in ammonia and other airborne pollutants that are a nuisance (odours), can contribute to acidification of terrestrial ecosystems, and are greenhouse gases. Liquid wastes high in nitrates can contaminate groundwater and reduce oxygen in surface waters, resulting in eutrophication; they also contain higher levels of heavy metals and phosphorus, which bio-accumulate from fertilisers and feeding supplements.

It is little wonder, then, that there should be growing environmental concerns about the *meatification* of global diets (Weiss 2007). The spread of factory farms across large parts of Asia and Latin America (having already become the norm across North America and Europe), designed to satisfy the rising urban demand of newly affluent households for meat, brings increasing pressure on land to produce feeds, and on the wider environment to cope with waste streams. However, drawing attention to such predicaments is not to make a moral claim for vegetarianism. We remain omnivores physiologically adapted to meat consumption; the question is whether the carnivorous appetites promoted by the food industry have distorted the balance in our food intake. Seeking an answer to this question is something to which we return in chapter seven.

6. Summary

This chapter sets out four major challenges for primary food production as we look toward the future. Climate change, freshwater depletion, "peak oil" and rising demand for meat represent the key issues that need to be addressed – and with some urgency – if we are to ensure stability and security of the global food system in the medium term. Dealing at some length with these four separate challenges is not to suggest that other issues are not also important. Maintaining the world's stock of plant and animal genetic diversity, comprising those varieties that have been carefully bred to produce food, as well as their wild relatives and the wider stock of biological resources, might for many represent the most pressing single issue that we face. Others would argue that the scale of land-use/land-cover change, particularly the conversion of tropical moist forests to arable farming or livestock grazing, represents the single most immediate problem.

The four issues identified here do not represent the only challenges that we face – and they cannot be resolved in isolation. In this regard, they help to demonstrate the need to embark upon a new course for primary food production that is less energy-intensive, more efficient in its use of water and other resources, and that results in lower greenhouse gas emissions and other forms of pollution. In short, primary food production in all its manifest forms around the world must endeavour to work within sustainable parameters. Chapter seven briefly sketches out what these might constitute. In the meantime, we need to examine the environmental consequences of the food system beyond primary production – the focus of chapter five.

Further reading

Hoekstra, A., Chapagain, A. (2008) *Globalisation of Water: Sharing the Planet's Freshwater Resources*. Oxford: Blackwell.
A vital sourcebook in aiding understanding of the significance of freshwater in international trade and why we need to address the disparities in the water footprint of nations.

Leggett, J. (2005) *Half Gone: Oil, Gas, Hot Air and the Global Energy Crisis*. London: Portobello Books.
As an ex-petroleum geologist, Greenpeace adviser and now CEO of a renewables energy company, Leggett is in a unique position to communicate a real understanding of the science and the possible consequences of both peak oil and climate change.

UNEP (2008) *Kick the Habit: A UN Guide to Climate Neutrality*. Nairobi: United Nations Environment Programme, www.unep. org/publications/ebooks/kick-the-habit/Pdfs.aspx
An extremely useful publication prepared for the UNEP-sponsored climate-neutral network. It contains a lot of material relevant to the food system as well as other sectors (energy, transport).

And on the web

Pew Center on Global Climate Change: www.pewclimate.org/global-warming-basics

"Climate change 101" provides an easy-to-understand primer on everything you need to know about anthropogenic global warming.

Water Footprint Network: www.waterfootprint.org

Association for the Study of Peak Oil & Gas (ASPO International): www.peakoil.net

The Meatrix: www.themeatrix.com
A brilliant set of online animation movies highlighting the problems of factory farming, with over 20 million hits so far.

5 Final foods and their consequences

Contents

1. **Introduction**
2. **Transforming foods**
    Social change
    Technological change
    The food industry
3. **The environmental dimensions of food transformation**
    Life-cycle assessment
    Refrigeration
    Packaging
    Life-cycle assessment: examples
    Cottage pie
    Tomato ketchup
4. **Transporting food**
    Extending the supply chain
    Road freight
    Air freight and food miles
5. **Food waste**
6. **Summary**
    Further reading

Boxes

5.1 The fractionation of corn
5.2 The development of the modern white sliced loaf
5.3 Water use
5.4 The global warming potential of refrigerants
5.5 The environmental impacts of a cola can

5.6 Bottled water: bad for people and the environment?
5.7 Evaluating the relative environmental effects of glass bottle reuse
5.8 Food air freight: chill chain via plane
5.9 Ecologies of scale in transport
5.10 Waste carrots
5.11 Luxus consumption

Figures

5.1 The co-evolutionary development of social, economic and technological change
5.2 Flowchart of life-cycle phases for a ready-meal cottage pie
5.3 Flowchart for Swedish tomato ketchup
5.4 CO_2 emissions by mode of transport, UK 2006
5.5 Changes in control over the food distribution system, UK

Tables

5.1 Environmental impacts related to the final consumption of products
5.2 Food transport indicators, UK 1992–2006
5.3 Illustrations of waste in the food chain

1. Introduction

Despite nutritional advice conveyed by the food pyramid that encourages us to eat more fresh fruit and vegetables, preferably in their raw or simply cooked state, much of the food that we actually consume is processed and made into a wide range of prepared foods and drinks. The overwhelming majority of food items filling the supermarket shelves are the products of a manufacturing industry that takes primary raw materials produced by agriculture – arable and animal – and turns them into a multitude of branded "final foods".

This chapter explores the stages beyond the farm gate – food processing, distribution and retailing – and some of their environmental consequences. Taken together with chapter two, it is an attempt to provide an overview of the food supply chain from field to fork, and to outline the multiplicity of ways in which food production activities have an impact upon the environment. Although there is not the space to provide detailed descriptions of processing technologies or distribution networks, this chapter provides evidence to suggest that the food industry has become one of the most innovative, dynamic and rapidly changing sectors. It is intensely cost-conscious and maintains considerable flexibility in sourcing its inputs in pursuit of achieving the slightest competitive advantage in the marketplace. It is also highly attuned to shifts in consumption behaviour – and plays a key role in shaping these changes – and is capable of responding quickly to an upsurge of interest in, say, healthy eating practices or buying more organic products. Finally, the sector is accustomed to technological innovation, traditionally in order to achieve improvements in productivity and cost savings, but increasingly this is accompanied by a need to demonstrate better environmental performance.

Over the past fifty years, the increasing globalisation of the food system has witnessed growing volumes of agricultural products, other raw materials and final processed foods criss-crossing the world in a manner that goes far beyond the earlier, and simpler, expediency of comparative advantage. The enmeshment of farmers in far-flung places to supply materials for a global industry appears to have partially "freed" food products from the constraints of nature and connections to place. Developments in transportation as well as in the molecular composition of foods have facilitated an increasing separation between places involved in the growing, processing, retailing and consumption of food. Extending the length of food chains has been immensely beneficial for these downstream interests – including the "non-discriminating" consumer – who can source cheaper materials on a year-round basis. It will become evident how significantly this process has been underpinned by cheap fossil fuels.

The first part of this chapter seeks to connect an array of interrelated social, economic and cultural changes with transformations in the form, content and availability of foods. Following this, the chapter explores the environmental dimensions of food processing and manufacturing. Here the notion of life-cycle assessment (LCA) is introduced as a framework through which to quantify the impacts associated with the many different food and drink products. Following discussion of refrigeration and packaging, this section closes by focusing upon two case studies designed to illustrate the extent and value of LCA. The issue of food transportation is examined next, and the notion of "food miles" critically evaluated. The chapter closes with a focus upon the issue of food waste, and explores the range of profligate practices within the modern food supply chain that contribute to a level of environmental impact that could be largely avoided.

2. Transforming foods

In the past, primary foods were generally subject to one of several well established preservation techniques in order to create robust, durable products for delayed or distant consumption. Today, such foods are more likely to be transformed and combined into an array of products with many different functional attributes for mass markets at national, continental and global scales. Such manufactured food products are not only expected to deliver durability and "shelf life", but are now designed to achieve a variety of other objectives. Depending upon the product, these might include time saving and convenience, nutritional value and health benefits, taste value, or any number of more quixotic claims such as "fun" and "adventure", terms frequently associated with food products aimed at children.

The twentieth century, particularly its second half, became the era not only of the industrialisation of agriculture, but of major technological transformations of food processing and retailing. As Lang and Heasman (2004: 139) remark, "New ways of packaging, distributing, selling, trading and cooking food were developed, all to entice the consumer to purchase." Advances in technology, involving the fractionation of primary foods into constituent elements; the development of synthetic additives (colorants, stabilisers, flavours and aromas) and their substitution for natural ingredients; and the emerging scale of production all combined to demonstrate a new era in the design, delivery and consumption of more processed and complex foods. These foods were the result of a highly dynamic process involving the interplay between societal changes, the development of industrial technologies and the growth of large food businesses capable of shaping, as well as responding to, changing consumer demand. These three elements are briefly described and explained below, their interactions demonstrated in Figure 5.1. While drawing broadly upon the UK experience, the discussion has wider relevance to other parts of the developed world.

Social change

From the mid-eighteenth century onward, the pace of social, economic and technological change increased. Changes on the land, not least the enclosing and privatisation of commons, forced many rural dwellers to find a livelihood in the towns and cities. Agricultural innovations, including the development of machinery, crop rotations and more systematic use of fertilisers, generated greater food surpluses that were required to feed the growing urban population. Yet, while industrialisation and the rise of the new manufacturing offered employment, conditions for the labouring masses were desperately unsanitary, with malnourishment and disease a generalised condition. Of the weekly budget of a typical poor household, two-thirds was spent on food, largely consisting of bread, potatoes, tea, sugar, meat, butter and beer – the latter three items being the first to be cut in bad years (Tannahill 1988).

By the mid-nineteenth century, the onset of the railways facilitated the provisioning of cities over greater distances and, in tandem with steam ships, helped transoceanic trade. Now, North American grain (and, later, Argentinean, Russian, Canadian and Australian wheat) all helped to ensure the security of bread supplies in Western Europe. The rise of refrigeration from the late 1870s led to a very rapid expansion of beef exports across the Atlantic. Refrigerated carcasses arrived in such volume that, by the early 1880s, cheaper American beef was already edging Scottish and Irish beef off the British market.

At the turn of the twentieth century, mass markets for food products remained very limited. A varied and nutritious diet was determined largely by wealth in the absence of an ability to produce food oneself, which, in the rapidly urbanising UK, was available to very few. For the middle class, grocery shops provided a range of preserved meats, tinned goods (meat, fish, fruit), and such factory-made processed foods as breakfast cereals, condensed milk and biscuits. The British working-class diet at the turn of the century was largely dominated by cheap cuts of meat, potatoes, white bread (eaten with jam or dripping), tea and sugar. A mass medical examination of 2.5 million men in 1917–18 found that 41 per cent were unfit for military service largely as a result of undernourishment. Circumstances did not improve substantially until the Second World War, when, food rationing was introduced across the UK and for the first time ensured that a nutritionally sound diet was available to all (Tansey and Worsley 1995).

The post-war period throughout the developed world witnessed profound social change that, to a significant extent, shifted the balance of power in society. While many of these changes were already under way in the United States, across Western Europe as well as in Canada, Australia and New Zealand a new more egalitarian mood took hold, with one of the most important developments being the greater number of women entering paid employment outside the home. This left fewer hours in the day to perform the many domestic responsibilities that traditionally lay on a woman's shoulders: shopping for and preparing primary foods to provide meals for her family being one of the most time-consuming. The emergence of the domestic appliances industry (also known as "white goods") was a response to this situation, designed to make available technologies that would free women's time from many daily chores (although not necessarily free the women from performing them).

Thus, during the 1950s and early 1960s, many households across the UK acquired their first refrigerator (usually following the acquisition of a twin-tub washing machine). While the latter helped with the drudgery of washing clothes, a refrigerator was able to reduce the daily, or at least twice-weekly, purchase of fresh foods (meat, butter, milk) that were otherwise kept in the cooler ambient temperature of the larder, simply a cupboard in the kitchen with a ventilation grill to the outside. The additional income from women's paid employment enabled households to acquire such consumer durables. Yet there was a second critical development: according to Goodman and Redclift (1991: 11), ". . . changes in food consumption, including the increase in 'convenience foods', probably would not have taken place at all without women playing the role of second wage earner". Although women continued to perform the traditional tasks of food preparation and cooking in the home, the nature of the foods prepared changed utterly. For example, in place of a dish such as a stew, requiring the peeling of potatoes and vegetables and long, slow cooking of cheaper cuts of meat, women could return home from work and heat through meal components from a tin, jar, or refrigerated or frozen packet. Inevitably this led to a reconfiguration of traditional dishes.

Many of the social, economic and technological changes that characterise this period from the 1950s onward demonstrate a deeply interconnected and interdependent set of processes that might be considered "co-evolutionary" in nature. Rising post-war incomes across the developed North offered households the opportunity to acquire consumer durables such as televisions and cars. Televisions became a critical gateway for advertising new food products as well as other consumer items. Although more recently television has struggled to maintain its advertising revenues, given competition from other media, it remains a powerful medium through which to shape food consumption behaviour. The irony of its grip on how we spend our time should not be lost, as a recent observation noted that all the time saved in the kitchen through the use of gadgets and the eating of convenience foods can now be spent in front of the television watching cookery programmes.

The automobile, too, has played a fundamental role in restructuring our food habits: first, in enabling the family to undertake the now weekly shopping at the one-stop supermarket, for which location decisions were less and less connected to pedestrian access or bus routes, but more to do with available space for car parking. Second, the car was instrumental in the development of the fast-food industry, as Schlosser (2001) so clearly describes: eating in cars probably had more consequences for the food than it had for the redesign of car interiors. Consequently, from both an ecological and a gastronomic point of view, the way in which these new systems of food provision emerged might be argued to have been designed principally to deliver convenience and low cost, but with consequences for the unsustainable and low-quality nature of the new food products.

Technological change

Primary foods in their natural raw state are biological matter, susceptible to decay and spoilage. This can occur through microbial activity – of

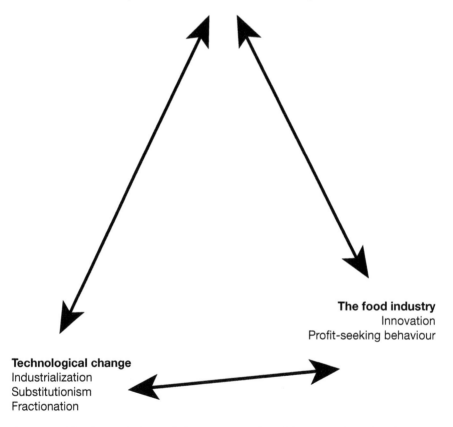

Figure 5.1 *The "co-evolutionary" development of social, economic and technological change*

bacteria, moulds and yeasts, some of which are pathogenic and can be extremely harmful to human health – chemical reactions and enzyme activities. Some foods are naturally more stable than others: nuts will generally keep for an extended period in suitable conditions without processing, while milk, on the other hand, will not. The first task of food processing therefore exists to slow down or prevent this process of decay. The long-established methods include drying, reducing ambient temperature, or employing one of a number of preservation techniques such as salting, smoking or pickling. Removing the product from air is another established method; the history of canning now stretches over two centuries and has long been joined by glass bottling and, more recently, the use of plastic films.

Many of the industrial food-processing activities that existed up to and even beyond the Second World War were based on technologies devised back in the nineteenth century, such as canning, steel-roller milling of grains, and sugar refining. There was evidence to support the argument that the food industry was marked by conservatism and an unwillingness to develop outside its core technological competence tied to specific types of primary inputs. Thus Nestlé, established in the 1860s, has spent much of its existence tied to milk products and confectionery, while a similar argument can be extended to other well known companies such as Kellogg's, Coca-Cola and Unilever.

However, in the post-war period, new food technologies emerged in association with the wider social and economic changes briefly outlined earlier. In part, this development was stimulated by the strategy of substitutionism, a process (outlined in chapter two) pursued by the food industry to reduce its dependence on specific raw materials, reduce their overall material and economic share in the final product, and thus create greater distance from the agricultural base (Wilkinson 2002). In place of specific primary ingredients required to manufacture a particular product, the food industry sought to develop greater interchangeability between a range of inputs, natural and synthetic. Consequently, the emphasis was on breaking down primary foods (fractionation) into a large number of constituent organic compounds – proteins, sugars, starches and alcohols – which could then be chemically modified, providing the basic building blocks for constructing a range of complex foods. Box 5.1 provides a description of the fractionation of maize that gives rise to a wide variety of food ingredients and industrial inputs.

Michael Pollan explains the principle of being able to substitute these different compounds, switching between maize or soy derivatives depending on their relative price, without altering the fundamental taste or appearance of the final product. This is the principle of "adding value": by so effectively disguising specific agricultural inputs, it is possible to reduce the prices paid to farmers such that for every dollar spent by consumers on cornflakes, 10 cents or less finds its way back to the growers; for corn sweeteners the farmer will be paid just 4 cents. It is the modern food industry's equivalent, argues Pollan, of some medieval alchemy capable of turning base metal into gold. Meanwhile,

---

Box 5.1

The fractionation of corn

The kernel of corn is first separated into its botanical parts – embryo, endosperm, fibre – and then into chemical parts. Each batch of corn is steeped in a bath of water containing a small amount of sulfuric acid, which swells the kernels and frees the starch from the proteins that surround it. The swollen kernels are then ground in a mill, with the germ taken off, dried, and squeezed for oil. Much corn oil is hydrogenated for use in margarine, in which atoms of hydrogen are forced into the fat molecules of the oil to make them solid at room temperature. These hydrogenated fats are now considered unhealthy.

The milling of the kernels results in a white mush called "mill starch"; this undergoes a progressively finer series of grindings, filterings and centrifuges. At each step, more freshwater is added: it takes approximately 22 litres to process 22 kg of corn. Wet milling is also energy-intensive: for every 1 kcal of processed food it produces, another 10 kcal of fossil fuel energy is burned. The end product is a fine, white powder: cornstarch. Although initially more popular in the laundry business (for whitening and stiffening collars), it soon became a major ingredient for food processors. By 1866, refiners had learned how to use acids to break down the cornstarch into glucose, a sweetener. As enzymes replaced acids, refiners were able to produce progressively sweeter sweeteners, and by the 1970s had perfected refining corn into fructose. The development of high-fructose corn syrup (HFCS) – a blend of 55 per cent fructose and 45 per cent glucose that tastes exactly as sweet as sucrose (from cane sugar) – is the single most valuable food product refined from corn, though there are dozens of other outputs for use in food and other industries. The fractionation of cornstarch yields adhesives, coatings and plastics for industry, stabilisers, thickeners and viscosity-control agents for food. Fermentation of mill starch creates a variety of alcohols, of which ethanol is becoming the most important as a source of fuel for automobiles.

Source: adapted from Pollan (2006), 87–9.

these base compounds, effectively stripped of their agricultural origins and created to perform a particular function, can be constituted within a new complicated contrivance with an extended shelf life. Under such circumstances, it is possible to understand better the vital role that the flavours and aromas arm of the food chemistry industry play in establishing the essential "food-ness" of the product.

Food chemistry has been an indispensable aspect of the technological changes within the food industry, contributing to the development of the most novel products as well as transforming the manufacturing processes of the most mundane. An excellent example of the latter is the white sliced loaf, which has become such a fixture on UK supermarket shelves that it has served as a loss leader for years. Box 5.2 captures some key aspects of the modern bread-making process.

Box 5.2

The development of the modern white sliced loaf

At the beginning of the nineteenth century, most bread was made in the home. But with increasing urbanisation and long working hours, the baking industry began to develop. In 1850, nearly all bakers were independents who baked bread and sold it on the same premises. By the beginning of the twentieth century, a few large factory bakeries were emerging, but the pace of change accelerated in 1932, with the arrival of a successful Canadian biscuit manufacturer who started buying up independent bakeries. The rush to integrate mills and bakeries thus began.

Until the 1960s, bread produced in factories was still made in a recognisable way, with dough left to ferment and prove over extended periods. But in the early 1960s, researchers at an establishment in Chorleywood developed a way of making bread that dispensed with traditional methods. Instead of allowing 2–3 hours' fermentation, they found that air and water could be incorporated into dough if it were mixed with intense energy at high speeds in mechanical mixers. Double the quantity of yeast was needed to make it rise; chemical oxidants were essential to get the gas in; and hardened fat had to be added to provide structure. But the process removed labour, reduced costs, and gave much higher yields of bread from each sack of flour because the dough absorbed so much water. Cheaper British wheat, with lower protein content than North American grains, could also be used. By 1965, the Chorleywood bread process (CBP) had become widespread.

The technology of the chemical improvers (or flour-treatment agents) on which modern bread depends has developed rapidly since. Improvers have long comprised hydrogenated fats, though with increased public concern over trans fats as a factor in heart disease, they are being replaced by "fractionated fat" derived from palm oil. The CBP also needs emulsifiers, which allow the bread to retain more air while slowing down staling, and these are commonly made from petrochemicals. High levels of yeast to speed fermentation, and salt to add flavour, are also needed.

Source: adapted from Lawrence (2004), 105–8.

The food industry

Through the development and application of advanced technologies that have facilitated the decomposition of primary materials to their molecular level, the food industry has arguably demonstrated a break with the conservatism of the past. In response to social change and technological possibilities, food manufacturing companies diversified, embarking upon a programme of mass product innovation. By the 1990s, tens of thousands of new product lines were emerging from factories to keep the growing length of supermarket food shelves stocked with their branded goods. Yet the competitive challenge faced by the food industry – not just horizontally with its rivals over market share, but increasingly downstream as multiple retailers became the real power-brokers of the food system – has required it to reinvent itself to a significant degree. Consequently, the largest and most dominant players within the sector are rebranding themselves as health companies, positioning themselves at the boundary with pharmaceuticals, particularly in the area of functional foods and nutriceuticals, where opportunities for adding value are at their greatest. Meanwhile, these new molecular-scale technologies enable companies effectively to perform the role of bio-refineries, utilising primary materials to produce a range of different applications such as bio-fuels (from milk whey as well as biomass), animal feeds, bio-plastics and other industrial materials.

As Pollan (2006: 96) explains, "you want to be selling something more than a commodity, something more like a service: novelty, convenience, status, fortification, lately even medicine". He uses the case of General Mills to illustrate this proposition. This US company started out in 1926 selling ground whole-wheat flour, but when that product became a cheap commodity as a result of competition, the company processed the grain a bit more, creating bleached, then "enriched" flour. This kept them ahead of the competition by selling not just wheat, but an idea of purity and health too. When even enriched white flour became a standard commodity, General Mills invented cake mixes and sweetened breakfast cereals. Today, writes Pollan, they are selling cereals that sound like medicines as part of the relentless pursuit of turning ever-cheaper agricultural commodities into more elaborate ways of adding value and inducing us to eat more brand-name convenience products such as their new line of "organic TV dinners".

Such innovation is driven partly by social change and technological possibilities, but also by the shifting balance of power within the agri-food system. For example, the growth of eating outside the home, the individualisation of eating as the composition of households began to change, the development of increasingly segmented markets differentiated by such variables as age, lifestyle, income, identity and so on, as well as the growing power of corporate retailing, all served to reshape the food industry. The growing sector of prepared foods (TV dinners) co-evolved with the widespread take-up of microwave ovens, but it also required the further development of cool-chain technology that would ensure refrigerated temperatures (of around 5°C) from the factory to the home. As Wilkinson (2002) notes, however, the "hollowing out of cooking activities" created by such developments also had to face the challenge of more radical substitution of the kitchen by the burgeoning sector of take-aways, as well as the extraordinary growth in ever more sophisticated snack foods.

The food industry, then, is deeply competitive, extraordinarily innovative, and commercially very successful. It is also marked by corporate avariciousness, as companies acquire manufacturing operations around the world through takeover or merger, extending their reach into new markets and building their portfolio of global brands. Yet while it appears to be dominated by some of the largest companies in the world, the food industry is actually composed of a large number of much smaller companies. Within the EU, the food industry is the leading sector in terms of production and employment, with 90 per cent of the total number of firms being small or medium-sized enterprises (SMEs). Many of these firms produce high-quality products for local and regional markets; it is important to recognise the diversity of the food industry and its many varied environmental consequences.

3. The environmental dimensions of food transformation

The manufacture of food (and drink) into tens of thousands of different products – some specific to local or national markets, others global brands consumed by millions around the world – encompasses many different processes with hugely varied environmental consequences.

A multitude of primary materials derived from agriculture or capture provide a vital stream of inputs: grains and cereals; fruit and vegetables; meat, fish, milk and eggs. These are the products of farms, plantations, ranches, concentrated animal feeding operations (CAFOs), trawlers and other operations. Some may be locally sourced, but a greater proportion are likely to involve significant transportation to the factory. All require fundamentally different technologies to transform them into edible products.

The sector comprises a host of different processing operations – for example, washing, drying, crushing, grinding, milling, slicing, mixing, moulding, cooking, extruding, fermenting and cooling. Those involving raising the temperature through process heating are likely to involve the on-site combustion of fuels (whether gas, liquid or solid) of fossil origin and to result in carbon dioxide emissions. Machinery used in processing operations – pumps, fans, conveyor belts, and those providing mechanical inputs – are likely to be powered by electricity which, unless there is on-site generation (possibly for back-up) will be drawn from the national grid. The environmental consequences here depend upon the type of generation.

In keeping with other manufacturing activities, food processing has generally grown in scale in order to achieve improvements in productivity (per unit of labour or unit of capital invested), increasing volumes of output and reducing (as far as possible) time spent in manufacture. Locations occupied by food-processing plants consequently have a significant environmental impact. For example, they require good transport access in order to take in large volumes of primary material from agriculture and essential supplies (chemicals, other food ingredients). They consume large amounts of energy (electricity from the grid; fossil fuels for heating and for motors, both static and mobile). In the UK, the food and drink industry accounted for 11 per cent of industrial energy consumption in 2002, excluding transportation (FISS 2006). The food sector is also a heavy user of water and a significant generator of wastes (solid, liquid, and as emissions to air). Box 5.3 provides a short account of water use within the food and drink industry.

The environmental impacts associated with the transformation of primary materials into final food products are not confined to the immediate vicinity of the manufacturing facility, however. Raw materials may be sourced from distant locations, and their production may leave a long-lasting environmental legacy resulting from the use of toxic agri-chemical inputs. The processed product must be packaged and labelled, requiring a quite different set of materials that can also originate in distant locations. Its onward movement, to retail stores or to food-service establishments, may require refrigerated transportation, and the cool chain may well extend up to the point of the product's final

Box 5.3

Water use

In food and drink manufacturing, water is used as a product (bottled), an ingredient (for example, in juices made from concentrate, in tinned vegetables), and most critically in a number of food processing activities, including washing, boiling, steaming and cleaning. The food and drink manufacturing industry is the largest water user in Europe and the USA. In Europe, it accounts for between 8 and 15 per cent of total industrial water use, and in the UK it is estimated to use 307 million m^3/year. This amounts to almost a quarter of the total water used by industry and commerce, and nearly 5 per cent of total water consumed in the UK. Compared with other industrial sectors, average water consumption per unit of product is ten to one, twice that of the chemical and five times that of the paper and textile industries.

A great deal of the water used by the food sector is for cleaning and hygiene purposes, and is consequently returned to the environment in a contaminated state. The dairy industry is the largest source of industrial effluent in Europe, with a typical European dairy generating a similar volume of effluent to that of milk processed (average 500 m^3/day). These effluents comprise by-products of processing (such as whey) and wastewater containing high concentrations of fat, as well as highly alkaline cleaning agents that present a significant challenge for treatment systems. Other food industries that present special challenges include breweries, with the second largest source of waste by volume, although its wastewaters are weakly contaminated; the distillery industry, which is lower in volume but very strongly contaminated; and meat slaughtering and processing establishments.

Given the enormous pressure upon freshwater resources arising from demographic growth, increased consumption and climate-change effects, as well as upon the capacity of the environment to absorb high-volume waste streams, there are efforts under way to improve the efficiency of water use in the food sector. These range from simple, low-cost measures to monitor water use more effectively (through metering and the use of more efficient nozzles and washing processes) to the installation of state-of-the-art wastewater treatment and recycling plant, involving advanced membrane filters, reverse osmosis and ultraviolet disinfection technology, and bioremediation.

Sources: adapted from Wheatley (2000); AEA (2007); CIAA (2007).

consumption. But, even then, the product exerts an environmental impact, with packaging often being disposed to landfill while the edible matter finds its own pathway back into the environment.

The purpose here is not to be indelicate, but to demonstrate the numerous environmental consequences associated with food. Because the accumulated impacts of any product – whether food or footwear, beverage or mobile phone – can be considerable, there has been some interest amongst researchers to develop tools that enable us to measure such impacts. It is important to recognise that this is not merely an academic exercise; on the contrary, if we are to find a way of reducing our collective burden on the Earth and to develop more sustainable societies, it is vital to have the evidence that demonstrates which products are problematic, or even if there are particular manufacturing process "hot spots" that can be improved. Consequently, analytical tools that enable a systemic and comprehensive approach capable of embracing the entire range of activities associated with a product or service have been developed and are being continuously improved and applied. This is known generically as life-cycle assessment (LCA).

Life-cycle assessment

An LCA approach represents a shift of focus from large point-source pollution activities (the factory and its atmospheric emissions, the landfill site and the failure of waste recycling) to one that looks at products themselves and attempts to understand their impacts at every stage in their life cycle. For this reason, LCA is often also referred to as a "cradle-to-grave" approach, providing a framework through which to evaluate and quantify all impacts from the extraction of raw materials, through manufacture and use, to eventual disposal as waste. There are four steps to undertaking an LCA. First, it is necessary to establish the goals and scope of the work so as to define very clearly the operating boundaries; second, to undertake an inventory of all relevant resources and emissions; third, to quantify the impacts; and finally, to make sense of what the data mean. Thus for a food product, a typical LCA would require a detailed analysis of the agricultural phase, measuring all the inputs (quantities of agri-chemicals, diesel use by machine, labour hours, etc.) as well as the output of primary product; a similarly detailed analysis of all subsequent processing/manufacturing activities; measurement of all movements of inputs and the product (distances and modes of transport being converted into volumes of diesel and petrol used); all the way to the product's purchase by the consumer, its storage, preparation and consumption.

This sounds exhaustive (and exhausting!) but there is more – not least, it is necessary to disaggregate just what we might mean by environmental impact for, clearly, all these different phases of the life cycle might give rise to quite different pollutants and forms of depletion and degradation. In order to be able to model the impacts (LCA generally requires the use of fairly sophisticated computer software underpinned by a huge data bank of conversion ratios), there is an understandable preoccupation with quantifiable parameters: things that can be measured. Some see this as one of the method's shortcomings, for it fails to account for more subjective issues such as the social implications of product use, while ecological criteria such as biodiversity or soil health have also been neglected (LCA was, after all, developed by engineers rather than ecologists).

The main environmental impact categories that have emerged as feasibly quantifiable and usefully comparable within an LCA include the following:

- climate change arising from emissions of greenhouse gases;
- eutrophication as a result of nitrate and phosphate emissions;
- photochemical ozone formation (smog) as a reference to low-level air quality;
- the effect of stratospheric ozone depletion as a result of the release of ozone-depleting substances;
- acidification (e.g. acid rain) associated with sulfur dioxide emissions;
- depletion of biotic and abiotic resources.

There appears to be less agreement over other impact categories, possibly due to difficulties in achieving a consensus over the appropriate scientific methods needed to measure them. These include human toxicity, eco-(aquatic and terrestrial) toxicity, and soil, water and land use.

An LCA of a particular food product consequently involves a multitude of measurements in order to derive some concrete statistical outputs that permit comparisons across categories and products. One example is the use of LCA to evaluate and compare the relative environmental impacts of an imported product against one produced domestically. Although one might assume that the additional transportation would add, for example, to the volume of greenhouse gas emissions, this may be offset by greater savings within the realm of production (fewer inputs). In this respect, LCA contributes to the evidence base as one of a number of useful analytical devices by which to appreciate the multiple environmental impacts of the entire food system.

Given the variety of food-transformation processes and the range of environmental issues that might be associated with each of them, it would be impossible and, frankly, a little tedious to attempt to describe the different food product categories and their problems. The next two sections focus upon two particular aspects of the food sector that raise significant environmental concerns: the increasing use of refrigeration, and the role of packaging. This is followed by a final review of LCA work through two case studies of everyday food products.

Refrigeration

Refrigeration is one of the most important elements in the food industry. It encompasses freezing, chilling, cooling or otherwise controlling the temperature so that food can be processed, ripened or matured in optimum conditions. Beyond slowing decomposition and therefore extending the life of products in storage and distribution, refrigeration temperatures improve the handling of certain foods ("sliceability", flavour extraction) and the solubility of CO_2 in carbonation. The principle of refrigeration is to remove the heat from the air occupying an enclosed space – and thereby also of the food stored in that space – whether domestic refrigerator or shipping container, and dump it to an external heat sink, usually ambient air. The conventional vapour-compression system requires the use of a volatile liquid as a refrigerant that is pumped around a closed system, cooling the cold plate and then shedding its acquired heat under low pressure as condensed water. The nature of this refrigerant has been a cause of global concern for more than thirty years.

Some of the longest used refrigerants were chlorofluorocarbons (CFCs), developed by the chemical industry from the 1930s and proven by science since the mid-1970s as responsible for the depletion of stratospheric ozone. This layer of ozone performs a critical function in screening out about 95 per cent of the Sun's harmful ultraviolet radiation, increased exposure to which can worsen sunburn, cause skin cancer and eye cataracts in humans, and lead to genetic mutations in plants, including food crops. International agreement to address this problem was embodied in the 1985 Vienna Convention, the resulting 1987 Montreal Protocol on Substances that Deplete the Ozone Layer, and subsequent strengthening amendments. With the development of CFC substitutes, hydrochlorofluorocarbons (HCFCs) and hydrofluorocarbons (HFCs), the chemical industry appeared to have responded to the urgent situation (having, for some commentators, done much to obstruct earlier international agreement) (Susskind 1994).

However, this second generation of refrigerants have proven themselves to be potent greenhouse gases with global warming potentials thousands of times greater than carbon dioxide (see Box 5.4). While there is now a concerted effort to develop a new generation of refrigerants that are neither ozone-depleting nor global warming agents, and hydrocarbons may perform this role satisfactorily, it demonstrates the dangers that can arise when new technologies, developed to provide a solution to one problem area, can lead to further difficulties elsewhere if there is insufficient precautionary evaluation. This has become particularly apparent given that refrigeration is now an indispensable and thoroughly integral part of the food industry.

In a review of the food refrigeration sector in the UK, Garnett (2007) has sought to calculate, based on the best available evidence, the energy used and greenhouse gas emissions arising from the operation of the cold-chain sector. Combining food manufacturing (*c.* 0.2 per cent), retail and catering (*c.* 1 per cent), and domestic (1.24 per cent) emissions, together with an estimate for leakages of refrigerants, she suggests a

Box 5.4

The global warming potential of refrigerants

Global warming potential (GWP) is a measure of the ability of a greenhouse gas to trap heat in the atmosphere. It reflects both its efficiency to absorb and radiate heat, and its lifespan in the atmosphere. As an index, it is compared on a unit basis with carbon dioxide, which has a GWP of 1. For example, according to the Fourth Assessment Report of the IPCC (2007), HFC-23, which is used as a refrigerant as well as a fire-suppression agent, has a GWP of 14,800 over a time horizon of 100 years, although it is estimated to be able to survive for 270 years in the atmosphere. HFC-134a, which is also widely used in automobile air conditioning, has an atmospheric lifetime of about fourteen years and a GWP of 1300. Its atmospheric concentration has risen sharply to over 10 parts per trillion by volume.

figure of around 3 per cent of the UK's total greenhouse gas emissions is derived from refrigeration. This does not, however, include the embedded energy of foods imported through cool chains from outside the UK, nor the energy consumed by mobile refrigeration units particularly as road-based goods vehicles.

It is as well to distinguish here between frozen goods and chilled foods. Frozen foods are usually blast-frozen to –25°C or colder and then generally stored and distributed at a temperature no higher than –15°C. Chilled foods are generally stored and distributed at 5°C, although in the UK the legislation permits a temperature of up to 8°C. A feature of the refrigerated foods sector in recent years has been a relative stagnation in the retail sector for frozen, and a considerable growth in chilled, foods. Two factors are major contributors here. First, the rise of ready-to-eat and convenience foods, including sandwiches with perishable fillings and other foods eaten on the go, many of which are designed to appeal to the health-conscious consumer (e.g. prepared salads). A second factor has seen the rise of chilled prepared meals. Garnett provides a useful illustration of what these changes involve: while consumption of potatoes in the UK has fallen steadily in volume since 1970, increasing shares of those that are eaten occur in processed form. While frozen chip sales remain static, there has been a significant growth in chilled potato dishes. This has significant implications not only for retail and domestic refrigeration, but also for transport, with increased energy use and greenhouse gas emissions.

Cool chains circumnavigate the globe, commencing within moments of picking in farms or plantations in tropical countries, and end only once the packaged produce is placed in a supermarket trolley. As a result of changing food consumption patterns, it appears that we are buying more highly perishable refrigeration-dependent fruits, vegetables, processed meats and dairy products. Such cool chains involve large-scale refrigerated transportation, involving trucks from packinghouse to dockside or airport, and again from point of disembarkation to cold store distribution centres.

Although refrigerated ships (known as "reefers") are responsible for moving large quantities of meat, fish, dairy produce, bananas, apples and other fruits around the world, a growing feature of cool-chain technology has seen increasing quantities of high-value perishable produce (e.g. cut flowers, strawberries, green beans and mangetout) transported by air and in a fraction of the time of surface transport.

Although the holds of reefers are divided up into spaces suitable for handling different foods, each requiring particular temperatures, the ongoing containerisation of shipping has led to large container ships offering greater flexibility and lower cost than specialist reefers. Individual containers are now equipped not only with their own refrigeration units, but also with the capability of modifying the atmosphere within to match the precise requirements of the foodstuffs. Controlled-atmosphere technology, for example, by reducing the presence of oxygen and increasing the proportion of other gases, is a way of extending the post-harvest life of perishable foods. The great advantage of containers is that they can be moved from ship to distribution centre without disturbing the contents.

The consolidation and resulting power and reach of the retail sector has enabled it to combine extraordinary logistical prowess – globally sourced, just-in-time continuous cool chains – with an ability to shape consumer demand for fresh and healthy (and therefore chilled) foods. This close association between the rise in corporate retailing and the growth in "food miles" is addressed in the next main section of this chapter. However, it appears that this process has also made us more refrigeration-dependent: the presence of refrigeration has shaped the kinds of foods we choose to eat. If we are to improve the environmental performance of the food system, then reducing this dependence on refrigeration will be necessary, which will mean choosing more local, seasonal and durable products, and fewer of those that are found in the freezers and cool cabinets of the supermarket.

Packaging

Packaging provides a number of important functions for food products:

- It protects and contains materials in liquid and solid form, providing some protection from crushing, or exposure to moisture, air or micro-organisms.
- It provides a means of communication, alerting the user to the brand and purpose of the food and its possible applications (including recipe suggestions); it also carries legally required information comprising quantity by weight or volume, the name and contact details of the producer, and nutritional information (composition of salt, sugar, fat. etc.), though this is not always in a form that is easily understood by most consumers.

● It performs a utility function, enabling retailers to stack and display possibly a large number of units on shelves, as well as for consumers in ease of handling.
● It carries a barcode that enables retailers to practise centralised stock-management systems, the scanning of products at supermarket checkouts registering in a central computer that triggers a re-ordering procedure as stocks run low.

Different types of packaging also perform different functions at different stages in the supply chain. Primary packaging (packets, bottles, cans) contain the individual product at point of sale to the consumer; secondary packaging (most often cardboard boxes or trays with clear plastic film) are the form used by retailers for stocking their shelves; while tertiary packaging (wooden pallets, shrink-wrapped, high-sided trolleys) provide protection during transport for distribution. In considering this topic here, most concern is with the primary packaging that ends up in the hands of the final user of its contents.

Primary food packaging accounts for about 17 per cent of municipal solid waste by weight, and between 20 and 30 per cent by volume within the EU. The food and drink industry accounts for about two-thirds of total EU packaging waste by weight (CIAA 2007). Packaging uses a number of different primary materials, including bauxite (to make aluminium cans; see Box 5.5), iron ore (to make "tin" cans from steel), silica (to make glass) and, above all, wood. It is estimated that half of all packaging consists of paper or paperboard, which is manufactured from wood pulp. This, in turn, is largely derived from forestry plantations of softwoods (invariably conifers), which generate their own environmental controversies. While the most frequently expressed criticism is that the practice of clear-cutting results in felled areas that are vulnerable to soil erosion, other concerns highlight the planting of single species monocultures that offer an unsuitable habitat for biodiversity, while such commercial plantations frequently detract from the visual amenity of a landscape. Transforming this wood into paper and paperboard packaging also has significant environmental impacts, though these vary according to the processes used. In general, mills employing mechanical pulping technology require large amounts of energy, while those using the chemical (kraft) process are associated with emissions of highly objectionable odours. Bleaching of the resulting pulp is one of the most controversial aspects of the manufacturing process, as traditionally it has involved the use of chlorine that gives rise to organochlorines such as dioxins, established carcinogens.

Box 5.5

The environmental impacts of a cola can

The can itself is more costly and complicated to manufacture than the beverage. Bauxite is mined in Australia and trucked to a chemical reduction mill, where a half-hour process purifies each tonne of bauxite into a half tonne of aluminium oxide. When enough of that is stockpiled, it is loaded on a giant ore carrier and sent to Sweden or Norway, where hydroelectric dams provide cheap electricity. After a month-long journey across two oceans, it usually sits at the smelter for as long as two months.

The smelter takes two hours to turn each half tonne of aluminium oxide into a quarter tonne of aluminium metal, in ingots 10 metres long. These are cured for two weeks before being shipped to roller mills in Sweden or Germany. There, each ingot is heated to nearly 900°F and rolled down to a thickness of an eighth of an inch. The resulting sheets are wrapped in 10-t coils and transported to a warehouse, and then to a cold rolling mill in the same or another country, where they are rolled tenfold thinner, ready for fabrication. The aluminium is then sent to England, where sheets are punched and formed into cans, which are then washed, dried, painted with a base coat, and then painted again with specific product information. The cans are then lacquered, flanged (they are still topless), sprayed inside with a protective coating to prevent the cola from corroding the can, and inspected.

The cans are then palletised, forklifted, and warehoused until needed. They are then shipped to the bottler, where they are washed and cleaned once more, then filled with water mixed with flavoured syrup, phosphorus, caffeine and carbon dioxide gas. The sugar is harvested from beet fields in France and undergoes trucking, milling, refining and shipping. The phosphorus comes from Idaho, where it is excavated from deep open-pit mines [. . .] The caffeine is shipped from a chemical manufacturer to the syrup manufacturer in England.

The filled cans are sealed with an aluminium "pop-top" lid at the rate of 1500 cans per minute, then inserted into cardboard cartons printed with matching colour and promotional schemes. The cartons are made of forest pulp that may have originated anywhere from Sweden or Siberia to the old-growth forests of British Columbia [. . .]

Palletised again, the cans are shipped to a regional distribution warehouse and shortly thereafter to a supermarket, where a typical can is purchased within three days. The consumer buys 12 ounces of the phosphorous-tinged, caffeine-impregnated, caramel-flavoured sugar water. Drinking the cola takes a few minutes; throwing the can away takes a second. In England, consumers discard 84 per cent of all cans, which means that the overall rate of aluminium waste, after counting production losses, is 88 per cent.

Source: adapted from Hawken *et al.* (1999), 49–50.

There is consequently some concern in finding ways to reduce the volume of traditional packaging materials, the processes that create them, and the raw materials used in their fabrication, without comprising the essential functions that packaging performs. This has led to a very rapid increase in the use of plastics, a generic term used to cover materials such as low- and high-density polyethylene, polypropylene, polystyrene, polyvinyl chloride (PVC) and polyethylene terephthalate (PET). Such materials have become utterly indispensable to the food and drink industry, providing relatively low-cost, light-weight, durable primary packaging options. One might argue that the extraordinary growth of the bottled water market could not have taken place without PET. In the USA, long since dominated by carbonated soft drinks traditionally packaged in glass, then aluminium, before the rise of PET, sales of bottled water have grown on average by around 8 per cent per year since 2000, exceeding $11 billion by 2008. With the exception of Germany, which has banned the use of PET for carbonated drinks, this material dominates the global market for water, a market worth around $70 billion. Indeed, so prevalent has bottled water become, that the country with the highest per capita consumption is Mexico, with 224 litres per person (Rodwan 2009).

Plastic packaging is derived from oil and natural gas. Through industrial chemical processes of distillation, fractionation, catalytic cracking and liquefaction, these hydrocarbons are broken down into a number of separate compounds, such as ethane, propane, butane, naphtha and gas oil. Ethane is the major feedstock for the manufacture of ethylene, propane is used to make propylene, while naphtha and gas oil yield benzene and paraxylene. These form the basic petrochemical building blocks for the production of packaging plastics:

- ethylene is used to make polyethylene, polystyrene, PET and PVC;
- propylene is used to make polypropylene;
- benzene is used to make polystyrene; and
- paraxylene is used to make PET (Selke 2000).

For those of us who are not industrial chemists, there comes a moment when we have to acknowledge the extraordinary ability demonstrated by a succession of scientists with a curiosity to devise such products from what do not appear to be encouraging primary materials. However, the fact that these packaging products are derived from a finite resource may become something of an issue for the food and beverage industry. This is not the place to speculate on whether the plastics industry of the future will switch to plant-based fibres with which to make its suite of polymers, and what the consequences of this might be. Rather, the final issue that needs to be addressed here concerns the issue of disposal of food and drink packaging, for the success of mass manufacturing – particularly of single servings of water and soft drinks – has been to create something of a waste problem. Thus, while the waste streams originating with consumer purchases of food and drink products are not exclusively made up of PET – linear low-density polyethylene originally made as supermarket carrier bags or polystyrene trays used to serve fast food are both often found as litter – it has comprised a growing volume.

Although PET (also referred to as PETE) bottles can be recycled, there are significant economic, behavioural and technical challenges that make it a less than optimal solution. It might possess its own resin identification code (#1) so that it can be sent for processing along its own recycling pathway, but in practice the low levels of consumer participation and regulatory incentives do not encourage a more refined segregation of different materials at source. The additional costs of recovery and sorting, and the likely contamination of the material, means that PET cannot be re-used for food and drink, but is broken down into a material known as PET flakes. These are then used to make polyester fibres that can be turned into carpets, bedding, clothing and so on. Yet barely 25 per cent of PET drinks bottles find their way into the recycling industry in the USA, meaning that 75 per cent, estimated at up to 140 billion items, enter the waste stream each year. It is no wonder that some people have begun to question the wisdom of drinking bottled water, and as they have done so, the mounting evidence points overwhelmingly to the illogical nature of the practice. Box 5.6 sets out some of the points made by critics of the bottled water industry.

In closing this discussion about the disposal of food packaging, it is as well to remind ourselves of the preferred rank order of managing waste (the 4Rs), as follows.

● *Reduction*: reducing the amount of material required to pack the product. For example, Nestlé claimed to have reduced the weight of packaging material used per litre of bottled water by 26 per cent between 2002 and 2006 (CIAA 2007). Unfortunately, its success in promoting its bottled waters has resulted in an equivalent increase in the number of units sold, thus making no overall improvement to the waste stream. The only serious solution here is for consumers to make the switch from bottled to tap water.

Box 5.6

Bottled water: bad for people and the environment?

Bottled water is being challenged and some of its claims to purity and health are being contested. Some points made by critics include the following:

- Despite labels claiming that bottles contain mineralised spring water, as many as 40 per cent of companies are simply purifying and bottling tap water.
- Tap water is more highly regulated than that found in bottles. In the USA, the Food and Drug Administration regulates only the 30–40 per cent of bottled water that crosses state lines, while the Environmental Protection Agency is responsible for continuous monitoring of public water supplies.
- A gallon of tap water costs US$0.02 compared with $8.26/gallon for bottled (a great deal more if one buys FIJI Water!).
- Coca Cola admitted in 2004 that Dasani water bottled in the UK came from the mains supply at its factory in Kent. The retail price represented a 3000 per cent mark-up over the cost of its raw material. Dasani made the company US$346 million in 2005.
- Nearly a quarter of all bottled water crosses national borders to reach consumers, transported by boat, train and truck. In 2004, Nord Water of Finland bottled and shipped 1.4 million bottles of Finnish tap water 4300 km from its bottling plant in Helsinki to Saudi Arabia.
- Making bottles to meet Americans' demand for bottled water requires anything between 1.5 and 18 million barrels of oil annually (conflicting data according to source), enough to fuel between 100,000 and 1 million cars for a year. Worldwide, some 2.7 million tonnes of plastic are used to bottle water each year.
- The US Conference of Mayors, at its 2008 meeting, passed a resolution to phase out the use of bottled water by municipalities. It recognised the irony of spending revenues on bottled water for employees and civic functions while promoting the quality of municipal water for its citizens. Municipalities also have to deal with a growing waste management problem.
- Only a quarter of used beverage containers are recycled, the rest – up to 140 billion per year – enter the waste stream for incineration or burial. Incinerating used bottles produces toxic emissions and ash; landfilled bottles can take up to 1000 years to biodegrade.
- Growing numbers of restaurants are now offering only tap water to diners.

Sources: adapted from BBC News (2004); Earth Policy Institute (2007); Food and Water Watch (2007)

- *Re-use*: returning the container to the food manufacturer for refilling (as was the case with milk and beer bottles; see Box 5.7).
- *Recycling*: the packaging material is broken down and remade, as in the case of glass bottles, which are crushed to cullet and used as feedstock in the manufacture of new glass bottles.
- *Recovery*: thermal processing (i.e. incineration) of packaging materials is effectively a means of disposal, though this feedstock may serve as a fuel for combined heat and power (CHP) installations. While this is preferable to disposal as landfill, it is the least satisfactory of the 4Rs.

In a book concerned with the environment, it is somewhat inevitable that an evaluation of food packaging would focus on the downside, on its degrading and depleting characteristics, rather than on its functional

---

Box 5.7

Evaluating the relative environmental effects of glass bottle re-use

Glass bottles are one of very few types of consumer packaging that have an extensive history of re-use. The environmental impacts of re-using glass packaging fall into three categories.

First, because bottles designed for re-use need to maintain strength for a longer time, they must be made stronger than one-way containers; therefore they are heavier and require more energy to manufacture – perhaps 50 per cent more. This increased weight will probably require an increase in the volume of secondary and tertiary packaging.

Second, the return system will require transportation of the containers from the site of use (doorstep milk delivery) or purchase (beer or carbonated drinks) to the refill location. This would be in place of the transport of new containers from the bottle manufacturers to the same location, and would save on the transport of those disposable bottles to recycling centres (or to landfill).

The third category concerns the cleaning necessary to prepare the containers for refill, involving water, detergents and sufficient energy to sterilise the bottles. One source used by Selke (2000) calculated that the energy required to fill non-returnable beer bottles was between 64 and 84 per cent of that for returnable bottles, while a second source reported that the energy requirements for a refillable beer bottle used 50 times were only 6 per cent of that for single-use bottles.

Source: adapted from Selke (2000).

strengths. It is clear that the packaging industry has made significant advances in improving its environmental performance, especially through reducing the volumes of materials employed. Efforts have also been made in reducing the presence of toxic materials, particularly as accumulated evidence emerges about the long-term health effects, for example associated with certain plastics such as phthalates, which have been linked to endocrine disruption; or PVC, which has been associated with cancer. A shifting array of external factors, including consumer preferences, environmental or food safety regulations, or market prices for primary materials, may cause food and drink manufacturers to switch between packaging options, for example using tetra-pak, a plastic-coated paperboard, rather than, say, PET. Finding ways of re-using packaging such as glass, perhaps associated with more localised food distribution networks, may also become a more necessary option in the future.

On the other hand, there is also the prospect of food packaging being at the cutting edge of the new nanotechnology paradigm, and this has the potential to utterly revolutionise the way we have viewed packaging in the past. In policy circles, nanotechnology is being heralded in much the same way as biotechnology was fifteen or so years ago, with the major food corporations working closely with university scientists to establish a range of lucrative applications. One of the most fruitful avenues to date has been in the realm of "smart packaging", where nano-sensors can be used to monitor the changing quality of the food contents and to detect the presence of pathogens or other contaminants. Other applications include the development of:

● barrier technologies to slow the decomposition process in food and thereby extend shelf life;
● a new range of nano-barcodes and monitoring devices, including nano-scale radio frequency identification tags able to track containers or individual food items – such nano-sensors would enable supermarkets to monitor product sales and expiry dates and facilitate product re-ordering;
● "smart fridges" for home use, designed to inhibit bacterial growth and eliminate odours, although there are concerns regarding the potential toxicity of nano-particles (Scrinis and Lyons 2007).

Food packaging is by no means a Cinderella of the food industry, and this brief discussion alerts us to the evolving challenges facing efforts to achieve a more sustainable agri-food system. But, in closing this section on food transformation, we return to LCA, and in doing so present some case studies that illustrate the complexity of environmental impacts.

Life-cycle assessment: examples

In this section we return to LCA as an approach that offers valuable insights into the multiple impacts of all kinds of products and services. The importance of this work is recognised by the European Commission as well as agencies of member states and other countries anxious to find ways of improving environmental performance. Many private companies, too, have supported such work, understanding the likelihood of future labelling requirements, the concerns of consumers, and the dangers of losing "first-mover advantage" to their competitors. In the case of the European Commission, a major study was commissioned in 2003 (Environmental Impact of Products, EIPRO), in order to draw together the findings of existing LCA research on the environmental performance of a wide range of products and services across the EU-25. The published report (Tukker *et al.* 2006) has been widely acknowledged as an important baseline from which to inform a process of integrated product policy, an approach that advocates life-cycle thinking and avoids policy measures that would simply shift environmental impacts from one phase to another.

Employing broad categories of consumer products and services, the grouping "food and drink, tobacco and narcotics" emerges as one of the most environmentally significant, responsible for 20 to 30 per cent of the various impacts of total consumption. Unsurprisingly, on disaggregating this category, the grouping "meat and meat products" has the greatest environmental impact, with a contribution to global warming within the range 4–12 per cent across the full production cycle, and a contribution to eutrophication of between 14 and 23 per cent. The second highest grouping is dairy products, with milk, cheese and butter contributing to global warming potential (4 per cent) and eutrophication (10–13 per cent). Other food product groups follow some way behind these first two and include, in descending order, edible fats and oils, bottled and canned soft drinks, bread, cakes and related products, then fruit and vegetables. Table 5.1 provides a summary of the type of environmental impact, the minimum number of product groupings responsible for contributing over half that impact, and a list of those food and drink product and service groupings that are included within that minimum number.

Table 5.1 demonstrates that of the eight products and services that make up more than half of total eutrophication within the EU-25, six are food and drink-related (the two that are not are "the driving of passenger

Table 5.1 *Environmental impacts related to the final consumption of products*

Environmental impact	Minimum number of products contributing*	Food and drink categories within minimum number of products†	Overall contribution of food and drink (per cent)
Abiotic depletion	7	1, 2, 3, 4	22
Eutrophication	8	1, 2, 3, 4, 5, 6	60
Global warming	11	1, 2, 3, 4, 5, 6	31
Acidification	12	1, 2, 3, 4, 5, 6, use of household refrigerators and freezers	31
Human toxicity	12	1, 2, 3, 4, 5, 6	26
Photochemical oxidation	12	1, 2, 3, 4, 5, 6	27
Ecotoxicity	15	1, 2, 3, 4, 5, 6, edible fats and oils	34
Ozone layer depletion	15	1, 2, 3, 4, 5	25

* Minimum number of products contributing >50 per cent of environmental impact.

† Food and drink categories: 1, eating and drinking places; 2, meat-packing plants; 3, poultry slaughtering and processing; 4, fluid milk; 5, sausages and other prepared meat products; 6, cheese (natural, processed and imitation).

Source: Tukker *et al.* (2006): 72–79 (constructed from data in tables 5.4.1a–h).

cars" (sixth) and "apparel (clothing) made from purchased materials" (eighth). It is little wonder that the category "food and beverages, tobacco and narcotics" accounts for 60 per cent of eutrophication in the EU-25. The contribution of this category to other environmental impacts is shown in the final column of Table 5.1. Once again, the high environmental impacts of livestock products is revealed, with the five items listed 2–6 in the key accounting for 16.4 per cent of the global warming potential derived from the consumption of all goods and services in the EU-25.

This gives an indication of how LCA can provide a general picture of the environmental consequences arising from the entire chain of food production and consumption activities. However, it is of particular value when applied to a specific product, enabling the researcher to construct an entire inventory of relevant resources and emissions and quantify their various impacts. In order to gain a brief insight into such an inventory, what follows is a highly summarised snapshot of two fairly mundane, everyday final food items: cottage pie and tomato ketchup.

Cottage pie

The UK Government agency Defra (Department for the Environment, Food and Rural Affairs) has embarked upon a process to develop, in collaboration with others, a Publicly Available Specification for the assessment of greenhouse gas emissions of goods and services throughout their entire life cycle. As part of this exercise, it commissioned research designed to assess the carbon footprint of a complex food (one constructed from various ingredients), in this case a cottage pie ready meal (Defra 2008b).

Notwithstanding the use of a range of environmental impact indicators as part of a full LCA, a great deal of work is currently under way that focuses exclusively upon the greenhouse gas emissions of a product's life cycle. Largely disregarding the measurement of eutrophication, acidification and so on, carbon footprinting adopts a cradle-to-grave approach in order to evaluate a particular product's contribution to global warming. As previously explained, other greenhouse gases, such as those derived from agricultural operations (N_2O, CH_4), can be converted to carbon-equivalent emissions. In this way, every life-cycle phase, from the manufacture of agri-inputs, the grazing of livestock, the transport of meat products, through to the energy used by the consumer's refrigerator and cooker, can all be evaluated and reduced to the quantity of carbon-equivalent released to the atmosphere. In this way, particular emissions hot spots are identified, ideally with a view to developing appropriate mitigation measures in order to reduce them.

In the case of our ready meal cottage pie, a flowchart of the life-cycle phases and individual components within them is shown in Figure 5.2. Although comprising twenty different ingredients, mashed potato and cooked (minced) beef make up over 70 per cent of the product by volume. Disaggregating the overall impact of the product, the phase responsible for the production of raw materials, which largely comprises agricultural activities, accounts for almost two-thirds of the global warming potential, while the other phases make quite modest contributions. Breaking down this raw material phase, in turn, reveals that while the production of potatoes contributes just 3 per cent of the greenhouse gas emissions, and milk products combined contribute 11 per cent, rearing the beef accounts for 69 per cent, or 45 per cent

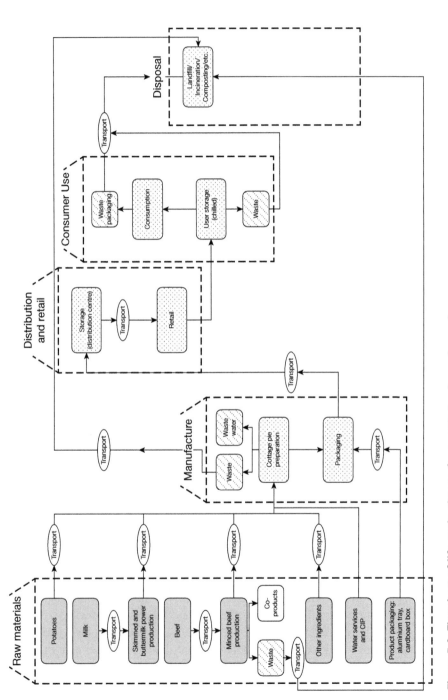

Figure 5.2 Flowchart of life-cycle phases for a cottage pie ready meal

Source: Defra (2008b).

of the total impact of the product. Consequently, again we find the primary production of meat the single biggest source of environmental impact (Defra 2008b).

Tomato ketchup

In one of the first studies of a processed food, Andersson *et al.* (1998) conducted a full LCA of a well known brand of tomato ketchup sold in Sweden. Their analysis involved six subsystems and a number of individual processes:

- *agriculture*: the cultivation of tomatoes and sugar beets, and the production of inputs to cultivation;
- *food processing*: the production of tomato paste, raw sugar, sugar solution, vinegar, spice emulsion, salt and ketchup;
- *packaging*: the production and transportation processes included in packaging systems for tomato paste and ketchup;
- *transportation*: all transportation processes except for the transports included in the packaging subsystem;
- *shopping*: transportation from retailer to household;
- *household*: storage of the ketchup bottle in the refrigerator, with calculations made for one month and one year.

The flow chart representing the life cycle of tomato ketchup is shown in Figure 5.3. Such a schematic diagram cannot truly convey the complexity of undertaking a scientific evaluation of a product's environmental performance, and some of the difficulties, assumptions, uncertainties and boundary-definition issues are outlined in their paper. It demonstrates, however, the value of engaging in such an endeavour, revealing the various environmental hot spots depending upon the particular impact parameter being considered. For example, we now know the agricultural subsystem represents a perennial hot spot for eutrophication, while domestic refrigerator leakage makes the household phase a significant contributor to ozone depletion. Yet, more surprisingly, the length of time that the bottle of tomato ketchup is stored in the refrigerator is a critical parameter of primary energy use of the product: storing it for one year uses as much energy as is utilised in the packaging and processing subsystems together!

This demonstrates that, while we have been pointing the finger at agricultural practices, especially those involving livestock, as being the primary causes of environmental impacts, the behaviour of consumers is also vital. Purchasing decisions about products are clearly significant,

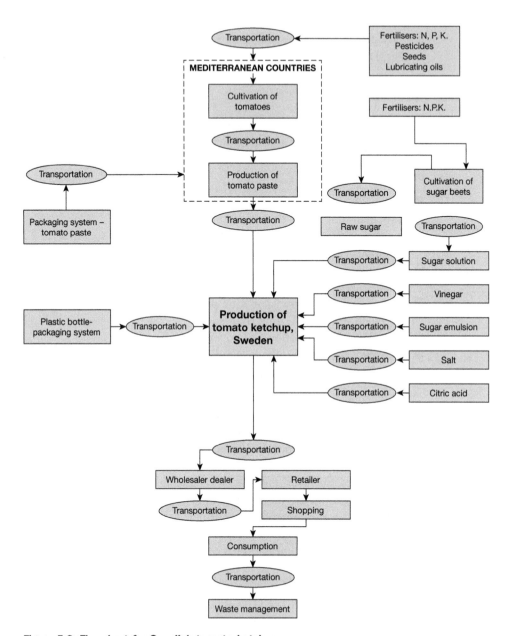

Figure 5.3 *Flowchart for Swedish tomato ketchup*

Source: Andersson *et al.* 1998, reproduced with permission of Elsevier.

but so is the way we shop for them (the use of a car, the distances travelled, the amount of groceries purchased); the way (and length of time) we store them; how they are prepared and how much we throw away. In the final section of this chapter we look at this aspect of the food system: food waste.

4. Transporting food

Perhaps, for the majority of people, supermarkets are simply a fact of life, a necessary evil or an indispensable feature of the modern world. We may give them little thought, though they have penetrated into the interstices of everyday life, in some places more than others. While in Europe corporate food retailers account for more than 50 per cent of all food retail sales, in the UK the top four retailers account for over 76 per cent and the top eight for over 91 per cent of all food and drink sales (Defra 2008a). If we were to reflect at all upon their existence, we might share the view that they are "an awesome phenomenon, a testament to the power of technology, logistics management and economies of scale" capable of channelling goods from an extensive global web of suppliers towards a nexus of retail outlets (Oram 2007). The corporate retailers who deploy the supermarket format of self-service in all its sizes and locations, as outlined in chapter two, have done more to shape food purchase and consumption practices in the North, and now increasingly in the middle-income countries of the South. These practices have rested largely upon an emphasis on convenience and low prices, though more recently this message has been modified to reflect a concern with quality assurance and traceability. Retailers are capable of promoting a powerful message of corporate social responsibility, partners in the development of local communities and, above all, key proponents of sustainable food systems. Yet, as will become clear, despite the claims of retailers to be essentially "green" companies with good environmental track records, large-scale food retailing and sustainability criteria appear to be operating in parallel universes.

As Morgan *et al.* (2006) observe, retailers have become the masters of managing space–time relations, and this is most evident in their development of the global fresh fruit and vegetable sector. A succession of innovations in supply-chain management has enabled retailers (and specialist third-party logistics companies contracted by retailers) to engage in arm's length control of primary producers so that they produce to the quality standards required at precisely the time needed.

Such producers are increasingly distant from the sites where their products are sold to consumers, a phenomenon that has given rise to the term "food miles". The purpose of this section of the chapter is to evaluate critically the way in which corporate retailers have shaped and extended their supply chains, and to examine the implications of this for modes of transportation and environmental impact.

Extending the supply chain

Reminding ourselves that changing consumer demands are as much a consequence of corporate food retailers' capacity to fashion them, as well as an ability to respond to them, one of the features of the past two decades or so has been the growing market for year-round fresh fruit and vegetables. In apparent consort with the public policy nutritional message for "five-a-day", corporate retailers have delivered a widening range of healthy, high-quality and convenient products. In any average supermarket in Western Europe or North America at any time of year, customers can expect to find pre-washed salad leaves; speciality vegetables (baby corn, mangetout peas, green beans); fresh, out-of-season fruits (strawberries, apples, pears, grapes); exotic fruits (mango, kiwi); as well as bouquets of cut flowers. Reading the labels of such products offers an insight into the globalised nature of the modern food system, with not only those countries we might expect (Brazil, Chile, Israel, Morocco) but others we might not (Burkina Faso, Mali and a host of others from Sub-Saharan Africa).

Amongst the citizens of northerly latitudes, there has long been demand for, and international trade in, commodities from warmer climes: bananas and citrus are long-standing cargo in refrigerated vessels, and later apples and grapes were shipped northward, filling counter-seasonal windows. However, trade liberalisation and advances in post-harvest technology and long-distance cold chains have driven rapid increases in trade in fresh produce. With an apparently strong consumer appetite for greater volumes of fresh produce, retailers embarked upon a strategy to develop this sector significantly, and not without reason: it carries some of the highest profit margins of any product category in store (Vorley 2003). The consolidation of the fresh fruit and vegetable supply chain in favour of retailers has consequently become one of the defining hallmarks of "supermarket power". While chapter six returns to this development and explore its implications for food security, here the focus is on the environmental consequences of this extended supply chain. For a more focused discussion the UK is considered in detail

Table 5.2 *Food transport indicators, UK 1992–2006*

Indicator	1992	1997	2002	2004	2006
UK urban food* (million km)	10,677	12,058	11,470	12,130	14,039
Index [1992 = 100]	100	113	107	114	131
Total HGV food† (million km)	7,862	8,412	8,121	9,104	8,965
Index [1992 = 100]	100	107	103	116	114
Overseas HGV (million km)	3,042	3,073	3,077	3,653	3,607
UK HGV (million km)	4,820	5,339	5,044	5,451	5,358
Air food (million km)	10	18	23	26	31
Index [1992 = 100]	100	186	236	269	321
CO_2 emissions (000 tonnes)	15,044	16,494	16,747	17,839	18,444
Index [1992 = 100]	100	110	111	119	123

* UK urban food combines car-based shopping, van and HGV food deliveries (urban centres only).

† Total HGV includes distances driven overseas to bring food to the UK as well as distances driven within the UK.

Source: adapted from Defra (2007).

here, largely because of the attention the topic of food transportation has received in this country; however, the issues raised have wide applicability throughout the global North.

In 2007, UK imports amounted to £25 billion and exports to £10.5 billion: a large volume of food and drink, all of which had to be transported into, out of and around the country. An estimated 91 per cent of imported food by weight arrives by ship: one quarter of this as bulk unprocessed commodities, the remaining three-quarters in containers or in refrigerated trailers. Steedman and Falk (2009) highlight the spatially concentrated handling of these imports, with three ports in south-east England (Felixstowe, Southampton, London) responsible for 63 per cent of container traffic, while Dover and the Channel Tunnel account for 41 per cent of truck-based traffic. This demonstrates that a greater part of the food chain within the UK begins in the south-east for onward distribution by heavy goods vehicles (HGVs) using the road network. Of course, this imported food has also travelled by road to sea ports for embarkation, or by truck across Europe. As regards air freight, while it accounts for only 1 per cent of total food transport, it has registered the fastest rate of growth of any, increasing on average by 9 per cent per year since 1992, and by 11 per cent in 2006. Table 5.2 provides the key transport indicators for the period 1992–2006.

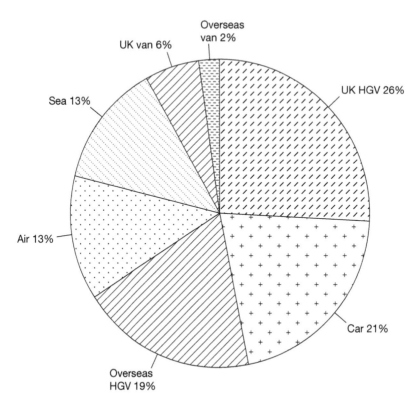

Figure 5.4 *CO$_2$ emissions by mode of transport, UK 2006*
Source: adapted from Defra 2007.

UK urban food kilometres increased by 6.7 per cent in 2006, the result of an 8.8 per cent increase in car-based shopping by consumers who appear to be travelling further and more often to visit the supermarket. In contrast, while total food imports into the UK increased by only 0.1 per cent in 2006 over 2005, air food kilometres increased by 11 per cent, bringing an additional 24,000 t of food. According to Defra (2007), this air freight was responsible for about the same amount of CO$_2$ emissions as the remainder arriving by sea. Food transport in the UK in 2006 was consequently responsible for 18.4 million tonnes of CO$_2$, of which 45 per cent can be attributed to HGVs. It is estimated that one-third of food transport via HGV is refrigerated; however, the emissions attributed are only from the engine for motive power, not from the diesel-driven refrigeration units, which are consequently not included in the pie chart in Figure 5.4. Clearly, freight transport by road and by air are two modes that need further elaboration here.

Road freight

First, the key to understanding the dominant role of road transport within the UK – as opposed to goods being put on rail or canals – has been the restructuring of the supply chains managed by corporate retailers. The extraordinary concentration of UK food retailing, with eight businesses accounting for over 90 per cent of sales, has enabled them to exert control over the entire distribution system, from field or factory gate to the supermarket checkout. This represents quite a shift in the way goods were traditionally moved. According to Steedman and Falk (2009), manufacturers were traditionally responsible for the delivery of goods to retailers, either directly to the store or to a company warehouse. Starting from the late 1970s, however, retailers began to develop their network of *regional distribution centres* through which goods were channelled, so that today over 85 per cent of products now pass through these centralised depots (with the exception of milk and bread, which are delivered to the store each morning by the suppliers). Although the retailers themselves took control of this secondary distribution between their own regional distribution centres and retail stores, this was extended further upstream. Figure 5.5 illustrates the stages of this development.

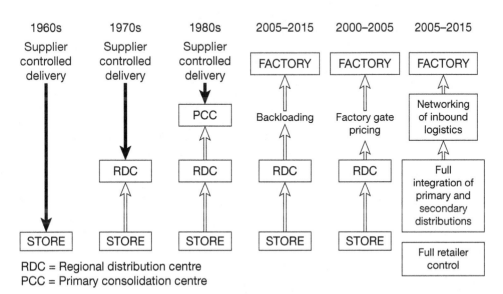

Figure 5.5 *Changes in control over the food distribution system, UK*
Source: Steedman and Falk (2009).

From the 1980s, *primary consolidation centres* were introduced to improve efficiency, particularly for fresh produce or for frozen goods, and are most frequently operated by a third-party logistics company on behalf of several different retailers. Incoming goods from a number of different suppliers are repacked and assembled into store-specific orders for onward transport via regional distribution centres. For other retailers, however, the drive to cut costs has encouraged them to take control of the entire distribution system, beginning with the collection of goods from the manufacturers. This has given rise to the phenomenon of "factory-gate pricing", whereby the manufacturer quotes the price of goods to the retailer with and without delivery costs, and the retailer decides which would be cheaper. This development, while demonstrating the growing hegemony of retailers within the entire distribution system, also illustrates the increasingly complex transport logistics of the food chain. For example, the transport of goods from a supplier to a regional distribution centre may be made by an HGV truck that had been used to make a delivery to a store in the area. Rather than returning to the regional distribution centre empty (or simply carrying empty pallets or packaging waste), it is scheduled to "backhaul" goods in the interests of efficiency.

The development of "just-in-time" systems, particularly for short-life products (fresh produce, dairy, meat), is designed to reduce to a minimum the amount of time that retailers are holding stock; the systems are designed to speed products through to point of sale. Naturally, with reduced stockholding the overheads are smaller; requiring small storage areas, lower risk of wastage and a higher rate of turnover, all in the interests of keeping down costs and improving profitability. Yet the environmental consequences are not so positive. As Steedman and Falk observe, the development of the retailers' national and regional distribution and consolidation centres and their associated networks has cemented the dominance of road travel. Not only that – despite the claims to great efficiency arising from all of the logistics, information technology innovations and efforts at backhauling, the distances travelled by food and drink on UK roads is increasing, while an estimated 30 per cent of total HGV movements on UK roads are of empty trucks returning to base.

Vehicle transport imposes various environmental, social and health costs, including those associated with congestion, noise, poor air quality (a particular problem for those susceptible to asthma and other respiratory ailments), accidents, infrastructure damage, and emissions

Box 5.8

Food air freight: chill chain via plane

About 80 per cent by volume of the air freight "perishables" sector comprises food and ornamental plants, though fish – edible but also ornamental – is a fast-growing component. Along with fresh produce, the sector includes prepared chilled products such as fruit salads and prepared vegetables and fish, and these value-added products are highlighted as particular growth areas.

Air freight export of food and flowers from many low-income countries, such as Kenya, Ethiopia and India, is growing. Tesco and Marks & Spencer reported an increase of 10 per cent in sales of Kenyan produce in the UK in 2007; and in May 2008, Asda announced that it would be sourcing an additional £30 million worth of fresh produce from African farmers over the next five years. The UK is a key target market; in September 2007, 60 per cent of shelf-ready products air freighted from Thailand to Europe were sold in UK supermarkets.

Perishable air freight infrastructure is widespread and prominent in airport expansion globally, with export facilities operational, in development or expanding in many countries, including Rwanda, Cameroon and Mali. Perishable produce accounts for approximately 80 per cent of air freighted exports from some African and South American countries. London Heathrow is a key airport for perishable imports, with British Airways (BA)'s cargo facility handling 115,000 t in 2006. Perishable imports are also flown into Gatwick and Stansted, with BA trucking produce to its Heathrow facility for national distribution. Kent Airport handled nearly 21,000 t in 2006, and Manchester Airport plans to increase its capacity to 18,000 t by 2010. Humberside Airport is a major gateway for fish, with imports from Iceland supporting processing in Grimsby. Imports to Robin Hood Airport in South Yorkshire include flowers from Florida, while exports include lobsters to Spain.

A complex chill chain extends around the globe. Agricultural produce is grown in the sun, but the chill chain might entail post-harvest blast chilling, on-farm chilled storage, a refrigerated truck to the airport perishable centre, refrigeration on the runway, and temperature control on the plane. Convoluted supply chains might include domestic flights to an export hub, interconnecting flights and multi-modal journeys. Some products labelled "produce of various countries" could involve multiple journeys. For example, in 2007 BA reported that mangoes were being shipped from India to South Africa for inclusion in prepared fruit salads, which are then air freighted to the UK for supply to supermarkets. After landing, the produce first goes to another airport perishable centre, then into refrigerated trucks to consolidation and distribution centres, then to the stores. This chill chain is often referred to as a 'weather-proof pipeline'; but, ironically, it is highly fossil fuel-dependent and a contributor to greenhouse gas emissions and climate chaos.

Source: adapted from Bridger (2008).

of greenhouse gases (primarily CO_2). Pretty *et al.* (2005) note that agri-food products accounted for 41.8 billion t-km, or around 28 per cent of the total freight transported in the UK, a proportion up from 25 per cent in 1980–82. (1 t-km is the movement of 1 tonne of food over 1 kilometre; rising t-km reflects an increase in volumes of food travelling greater distances.) These authors then calculate the total external cost of this transport, that is, the hidden costs associated with such problems as those noted above, which are not factored into the bill paid by the consumer in the purchase of the weekly food basket. Based on these calculations, Pretty and colleagues estimate that UK consumers are paying around £39 per person per year in additional environmental costs associated with road-based freight transport to retail outlets, the largest external cost element associated with the movement of food from farm to table.

Air freight and food miles

This brings us to the second problematic mode of food transportation: air freight. Although it will be recalled from Table 5.2 just how rapidly air freight has grown in recent years, Box 5.8, drawn from a paper by Rose Bridger, provides an extraordinary insight into the way it continues to expand worldwide. It is clear that air freight-dependent supply-chain management anticipates a future of growing demand in Europe for African horticultural products, Thai shrimp, Icelandic cod and Indian ornamental fish. But why should this be a problem?

Aircraft not only produce CO_2 from the combustion of jet fuel and other atmospheric emissions including nitrogen oxides, sulfur oxides and soot, but they do so directly into the upper troposphere and lower stratosphere, where they have an impact on atmospheric composition. For example, the condensation trails left by aircraft, most clearly seen in a clear sky and known as "contrails", are thought to trigger the formation of cirrus cloud which acts as a further "warming blanket" around the Earth. Together with the non-CO_2 gases, which have shorter atmospheric residence times, there is evidence to suggest enhanced radiative forcing within the mid-latitudes of the northern hemisphere, where there is a particular concentration of aircraft movements. Although the atmospheric chemistry involved is extremely complex (the emission of nitrogen oxides by aircraft in the lower stratosphere is thought to have a role in ozone formation) and well beyond the immediate needs of this text, it is clear that more rather than fewer flights are likely to exacerbate rather than mitigate environmental changes such as global warming.

Which brings us, inevitably, to the question of food miles. This is a label that has achieved considerable currency in popular discourse, a success borne of its relative simplicity: the greater the distance travelled by food, the less environmentally "sound" the product might be. Yet, as we have seen, the primary production, transformation and distribution of food is a complex business with a multitude of environmental consequences that go well beyond a singular aggregate measure of miles (or kilometres) clocked up. Production methods, modes of transport, resources used, amongst other considerations, all have a bearing on the overall environmental performance of the product. Counter-seasonal supply of fresh produce, for example, might carry high food miles, but if the alternative is to extend the growing season using heated greenhouses, it invariably emerges with a lower global warming potential. Edwards-Jones *et al.* (2009) review research undertaken into various horticultural supply chains, and observe the following:

● The production and air freight of green beans from East Africa to the UK results in a global warming potential (GWP) five to six times greater than seasonally grown beans in the UK, and is dominated by the air transportation stage.
● The growing and refrigerated road transport of lettuce from Spain to the UK market during the winter season has a lower GWP than growing lettuce in protected environments in the UK.
● In the case of broccoli, a comparison of supply chains from UK and Spanish production revealed that fresh UK product had the lower carbon footprint, but if frozen its footprint was higher than fresh Spanish broccoli transported to the UK. In all cases, the life-cycle hot spot for broccoli was in home cooking, accounting for between 50 and 70 per cent of greenhouse gas emissions.

It is apparent, then, that making basic assumptions about the fewer the food miles the better can be deeply misleading in terms of environmental impacts, and can also detract from appreciating the wider social and economic implications. For example, the fresh fruit and vegetable sector in East Africa supports the livelihoods of at least a quarter of a million people employed in field and packing tasks. A proposition to reduce food miles and end the import of green beans from Kenya would immediately jeopardise the wellbeing of these workers and their families (an issue to which we return in chapter six). On the other

---

Box 5.9

Ecologies of scale in transport

Approximate amounts of CO_2 associated with moving 1 tonne (t) of goods over 10 km (in grams):

- on a fully loaded, 10,000 dead weight t freighter: 200
- as part of a 20 t load on a 40 t truck: 400
- in a van carrying 1 t: 2400
- in fifty cars each carrying 20 kg: 10,000

Source: Foster (2011) personal communication.

hand, the term "food miles" permits an opening through which to raise the question: what are the environmental limits of consumer sovereignty? To what extent is it reasonable to continue to assume that air-freighted strawberries should be available in northern hemisphere supermarkets all year round? Moreover, what are the aggregate consequences of car-based supermarket shopping trips, the dominant model through which households in the North procure domestic food needs? This last question is addressed by a simple but highly effective table by Chris Foster, shown in Box 5.9.

Although proportions will clearly vary between countries, in the UK, car-based supermarket shopping trips account for 48 per cent of total food vehicle kilometres, and 13 per cent of CO_2 emissions associated with UK food transport. Moreover, as noted above, this sector is growing at a faster rate than that of total urban food transport, rising by 8.8 per cent in 2006. This trend appears to reflect the rise in the number of out-of-town shopping centres, featuring "mega-stores" and hypermarkets, at the expense of city-centre supermarkets and independent grocery shops. Clearly, the key to this development is scale: the provision of free car parking adjacent to the store facilitates the purchase of more things, in larger units, taking advantage of special offers ("buy one, get one free"), all of which present opportunities for savings, real or imagined. It also results, as we see below, in generating more potential waste. For retailers, larger stores frequented by more customers who buy more things represent a faster turnover of products and an opportunity to increase profitability. The fact that the private gain of shoppers and retailers is achieved at a high collective social cost – one estimate is that car-based shopping is responsible for 40 per cent of the total external costs of £9 billion per year for food transport (Defra 2005) – demonstrates the environmentally short-sighted nature of this model.

Although we have focused in this section upon the way in which corporate retailers have shaped and extended their supply chains and the consequences this has had for transportation and distances travelled, this is not the only environmental impact associated with retail operations. Leaving to one side a host of locational impacts and direct emissions to air and water arising from the operation of a supermarket store, there are three key areas that have been identified as climate change hot spots. The first concerns temperature-controlled products that evidently have a much greater energy demand than those accustomed to ambient temperatures. It has been estimated that up to 60 per cent of a supermarket's electricity consumption is used to drive refrigeration equipment, with the remainder used for lighting, heating, ventilation and air conditioning, as well as in-store baking and other services. Although stores draw heavily from national electricity grids in order to meet their considerable energy demands, the composition of their indirect emissions will reflect the mix of power-generation technologies deployed by the grid.

Secondly, refrigeration systems in supermarkets contain large amounts of refrigerants and, even with the use of sensing equipment, a leakage rate of 15 per cent per year can still account for up to 30 per cent of indirect emissions of greenhouse gases. It has been estimated that energy consumption of refrigeration display cabinets and refrigerant leakage account for over 80 per cent of greenhouse gas emissions from the distribution and retail phase of the food product life cycle. Third is the issue of food waste, which in the UK retail sector has been estimated to amount to 1.6 mt per year, the vast majority of which is disposed to landfill. Which brings us, appropriately, to the final component of the agri-food chain.

5. Food waste

Food waste is a troubling issue that goes well beyond its contribution to climate change through the release of methane from decomposition in landfill sites. For those of us raised by parents with direct experience of wartime rationing, childhood mealtimes that didn't involve finishing the food on your plate were likely to bring forth admonishment and a reminder of the "starving children in Africa". More recently, Stuart (2009) has sought to establish the connection between food profligacy in rich countries and food poverty elsewhere in the world: if the former buy hundreds of millions of tonnes of food and end up throwing it away, these countries are effectively and gratuitously removing food from the market that could have been bought by others. As we see in chapter six, the availability of food does not necessarily mean that people are able to access it, but the point is well made.

Wasting food is therefore a moral issue, but it is also an intensely environmental issue. The good news for children everywhere who have yet to see the joys of cooked vegetables is that the leftovers on their plate actually constitute only a fraction of total food waste; the bad news for all of us is that the modern food system, from the field to the table, has become characterised by a scandalous level of discard and wastage. Table 5.3 provides a breakdown of the kinds of food losses and waste that can occur at each stage of the agri-food system in its most generic form, taking in all types of food chains wherever they occur in the world.

At the first stage of the food chain, at field level, losses can occur as a consequence of circumstances beyond the capability of farmers to prevent them, with many developing countries struggling to reduce crop losses due to a range of environmental hazards as well as pests and diseases. For example, while conducting research in the early to mid-1990s in southern Sumatra, Indonesia, in an area recently settled by

Table 5.3 *Illustrations of waste in the food chain*

Stage	Examples of food waste and loss
Crop growth and harvest	Loss to pests (insects, birds, rodents); edible crops left in field due to rejection by retailers ("outgrading" stringent quality standards or surplus to contract); losses in harvesting (bruising)
Post-harvest processing (threshing, drying), storage, transportation	Spoilage, spillage, contamination; discards from quality control
Processing and manufacture (final foods); packaging and distribution	Process losses/trimmings; discards from production over-runs, quality control, damage in transit; equipment failure (refrigeration)
Retail	Discards from surpluses from overstocking, best-before dates, packaging damage, blemish
Food service sector	Food-to-go operations: disposal of unsold stock at end of day; sandwich shops: crusts and end slices
Consumers	Poor stock management in home: discards of best-before items, BOGOF purchases; edible elements for aesthetic reasons; prepared food surplus to servings; plate scrapings; over-eating

Source: adapted from Parfitt et al. (2010).

migrants, I was struck by the enormous challenge of crop depredation by wildlife and rodents. Although the elephants and monkeys had largely disappeared from the area as their habitat gave way to a farmed landscape, wild pigs remained a serious nuisance, and rats an even greater problem. In one of the villages in which I worked, farmers reported that up to 95 per cent of their rice harvest in 1993 was lost to rats, leaving them exposed to extreme hunger.

In industrialised farming systems with highly controlled production environments, losses to pests and diseases are generally minimal given the stringent measures that are normally available, but weather remains a source of uncertainty for farmers, with late frosts, high winds, hail or excessive rain the cause of crop losses. Yet the bigger problem for producers contracted to supply a large retailer with a given volume of a crop is the imposition of quality criteria that result in high rates of outgrades or even the complete rejection of a crop. In Box 5.10, Stuart (2009) describes his visit to a Yorkshire carrot grower, where there appears to be a high and unnecessary level of food waste as a consequence of cosmetic standards.

Drawing on her observations, Lawrence (2004: 133) asserts that "(t)he supermarket 'grade out' has become a source of great dread to farmers around the world. Enormous quantities of food go to waste because they do not meet the very narrow specifications now demanded by most big supermarkets." She describes a fruit grower in Cambridgeshire who grows thirty-two varieties of plum that can ensure a continuous supply of ripe fruit from mid-July to the end of October. But the supermarkets take only three varieties for a few weeks and the specifications are strict:

> They've got to be thirty-eight millimetres, unmarked, with stalk, to pass muster . . . The shape must be "typical", they must be more than fifty percent coloured but they've also got to have a four-day shelf life. Well, that means they've got to be picked rock-hard. Plums need to be picked and eaten within a day or two to taste good.
>
> (Lawrence 2004: 132–33)

It seems that most of the fruits that fail to meet these specifications are thrown away. But a similar process occurs in Kenya, where smallholders grade green beans for export to UK supermarkets:

> They stood in a simple traditional building . . . measuring the vegetables against wooden slats to check that they conformed to the requirement that a green bean should be 95 mm in length and between 5 and 7.5 mm in diameter. And they had to be straight; curved green

beans would not do. The packer and exporter, Homegrown, explained that 35 percent of beans failed to make the supermarket grade, and while a few of the rejects went to cattle feed, or into the local market, most went to waste, even in a country where people go hungry.

(Lawrence 2004: 133–34)

Box 5.10

Waste carrots

Guy runs a carrot farm and packing plant that handles 70,000 t of carrots per year, and supplies Asda, the Walmart subsidiary in the UK. Once the harvested carrots are washed, they then pass through a pair of sensor machines that check each carrot for colour, straightness and blemishes. Those that pass continue onto a conveyor, where operatives grade and pack the carrots into plastic bags for the supermarket. Nearly one-third of all the carrots are outgraded: half of these are rejected for physical or aesthetic defects, being the wrong shape or size, including not being perfectly straight. Guy is able to sell about one-third of these outgrades – the oversized specimens – to food processors, but the rest are consigned to livestock feed, which recuperates barely 10 per cent of their value. Though these carrots are perfectly good for human consumption, a series of "quality controls" prevent them from entering the food chain. At the European level, there are strict specifications on the size and shape of fresh produce in order to conform with its designated Classes. Yet retailers generally operate even more stringent standards, although they like to claim they are responding to the fastidiousness of consumers.

Fearful of being delisted as a supplier, Guy ensures that he meets the requirements of the retailer by planting 25 per cent more carrots than he needs as a buffer to cover any eventuality. This is common practice in the field vegetables sector, where planting 140 per cent of actual demand is normal. And what does Guy do with the surplus? Some he will harvest and sell to wholesalers and processors, but the price differential with what he receives from the retailer makes secondary sales unattractive. Consequently, each year Guy ploughs into the soil 10–12.5 per cent of his crop. Together with the outgrades, less the sales of oversized roots to processors, this means that around 28 per cent of Guy's carrots fail to make it into the human food chain. Adding together losses between the farm and the consumer, together with what is wasted by consumers, this means that of all the (commercially grown) carrots in the UK, 58 per cent ends up as waste. For every carrot purchased in a supermarket, the consumer has paid for at least one more to be thrown away, while fresh salad is even worse: for every serving eaten, two more have been thrown away.

Source: adapted from Stuart (2009).

Significant volumes of food waste also occur at the food-processing and manufacturing stage, although it is apparent that precise figures do not yet exist. In the United States, estimates suggest that around 10 per cent of raw materials entering food-processing factories is wasted, while mass balance studies in the UK indicate that around 16 per cent of materials entering food and drink industries is discarded as waste (Parfitt *et al.* 2010). Clearly, there are likely to be enormous differences in the volume and type of waste generated by different food-processing activities, with some likely to produce unavoidable by-products (animal tissue, inedible plant material), while others are guilty of discarding perfectly good food. In those sectors producing ready-to-eat convenience foods within the chill chain, there are likely to be much higher rates of wastage than those producing more durable ingredients such as tinned or packet foods held at ambient temperatures. Stuart (2009) reports on a company producing packaged sandwiches for Marks & Spencer that is required to discard the crust and the first slice at either end of every loaf of bread it uses, amounting to 17 per cent of each loaf, or 13,000 slices from a single factory every day. In addition, there is overproduction waste when supermarkets revise their higher forecast order, leaving suppliers with pallets of fresh sandwiches packed into the retailers' dedicated livery packaging and therefore unsuitable for sale elsewhere. Such food invariably finds its way into landfill.

Finally, to the point of consumption, where individuals and households make decisions about the food and drink they wish to consume, how and where they purchase it, and what is thrown away. According to European Commission data for the EU-27, while household consumption expenditure increased by 29.3 per cent between 1995 and 2006, expenditure on food rose by just 15 per cent, indicating the general downward trend in food's share of total household expenditure. Interestingly, spending on non-alcoholic beverages (bottled water) rose at a slightly faster rate (31.7 per cent) than total household spending, while expenditure on catering services rose at almost 25 per cent, illustrating the degree to which eating outside the home has become an increasingly important aspect of food and drink consumption (Eurostat 2008).

Food wastes arise at the site of provision of "eat-in" food by canteens and restaurants, as well as at various sites where take-away and convenience food and drink products are consumed, either on the move (dashboard dining) or at home (the Indian or Chinese take-away). The total waste stream in the take-away category generally comprises both an organic fraction (food residues) as well as the array of packaging in which it was supplied – much of which is non-biodegradable (polystyrene, aluminium foil). This take-away debris frequently represents the most visible aspect of food waste, often as litter. However, it is on foods purchased for preparation and consumption at home that most attention is focused, and where the greatest concerns arise.

Before considering estimated quantities of waste, it is as well to remember that the *massification* of food product manufacture across the developed world has resulted in a historically unprecedented low price for dietary energy. Moreover, with the food system built around the rapid throughput of large volumes and profitability contingent upon unit sales, every incentive exists for food service providers and retailers to encourage shoppers to buy more than they need. In the fast food sector, this phenomenon was captured by Morgan Spurlock's film *Super Size Me*, the title reflecting the invitation to customers to choose the largest serving of the selected menu item at little additional price. Its equivalent in food retail sales is the BOGOF promotion: "buy one get one free". Such deals encourage consumers to buy, to eat and to throw away more than they might otherwise do.

Studies designed to quantify levels of household food waste are complex, not least in distinguishing between "edible" and "inedible" waste fractions (crusts of bread versus used tea bags may be straightforward, but what about potato peelings?), and also in the way these are disposed of (wastes to drains, scraps fed to pets, backdoor composting, the municipal waste stream). Clearly there are many factors that contribute to waste: household size, composition and stage in the life cycle (are children fed exclusively "kids' food" or do they share the common family meal?); cultural attitudes such as notions of hospitality that encourage the preparation of more than enough food; income level (lower-income households generally waste less than the better-off), and so on. Although some food is wasted as a result of preparing too much (left in the pot or on the plate), a great deal of waste arises simply from poor stock management and buying too much.

One issue that has been identified in this regard is the confusion surrounding various date labels on food products, in particular the lack of distinction between "best before" and "use by" dates. Generally, best before dates are applied to lower-risk food – such as baked products, those stable at room temperature, and chilled foods likely to be properly stored – and simply indicates a date up to which a food would be at its best, but would not be unsafe after it. Use by dates, on the other hand, are applied to pre-packed, perishable foods that could become unsafe to eat after the date, and include cooked and partly cooked or cured products largely derived from livestock or fish. Supermarkets using their own dating systems for stock management purposes, and applying these to fruit, vegetables and bakery products, have exacerbated consumer confusion. As a consequence, a large amount of perfectly safe and edible food that has reached its best before date is thrown away under the impression that it has "gone off". Hence, in the UK, "484 million unopened pots of yoghurt, 1.6 billion untouched apples and bananas worth £370 million (Stuart 2009: 71) end up in landfill.

Making international comparisons in the levels of household food waste is necessarily fraught with difficulty, given differences in the methodologies used in each country as well as some of the issues noted above. Nevertheless, Parfitt *et al.* (2010) draw upon a number of studies that demonstrate quite a wide range in levels of food waste. In the USA, a study by the Environmental Protection Agency estimated that food waste was around 233 kg per household per year, amounting to almost 32 million tonnes in total and accounting for almost 13 per cent of the municipal solid waste stream. In the Netherlands, Dutch consumers are estimated to throw away between 8 and 11 per cent of food purchased, equating to between 43 and 60 kg of food waste per person per year. In South Korea, the introduction in 2002 of a ban on food waste entering landfill occurred when food accounted for 27 per cent of household waste. "Despite an awareness-raising effort in advance of the ban, food waste increased by almost 6 percent over 4 years after the ban, with increased consumption of fresh fruit and vegetables linked to higher incomes cited as a reason" (Parfitt *et al.* 2010: 3074). Finally, in the UK, estimates of household food waste range between 240 and 270 kg per year, perhaps representing as much as 25 per cent of that purchased by weight.

Why do all these estimates represent an unacceptable level of food waste? Because, in all cases, they represent a serious failure to acknowledge and account for the resources and ecological services utilised and embedded in that waste food: whether soil nutrients, freshwater, fossil energy or animal life. Responsibility for food waste might be attributed to the usual suspects – big business or the failure of governments to regulate adequately – but ultimately it rests with our own shortcomings to avoid the allure of purchasing more than we need.

One final illustration of the way in which food might arguably be regarded as wasted is through individual excess consumption. As noted previously, the rationale of the modern food system is to maximise the throughput of products such that food purchases in excess of what is actually required for a sufficient diet may end up either going directly to landfill (the unopened yoghurt reaching its best before date in a domestic refrigerator), or are eaten and contribute to rising levels of body fat. Arguably, then, overeating represents a means of disposing of "surplus" food, but with important consequences for personal health and wellbeing. Some relevant observations made by Blair and Sobal (2006) in this regard are outlined in Box 5.11. As they suggest, as concerns

Box 5.11

Luxus consumption

In their paper, Blair and Sobal (2006) use the term "luxus consumption" to refer to the acquisition, consumption and discarding of food beyond physical needs. Calories ingested beyond the fuel needs of the body and not dissipated by metabolic processes result in body fat accumulation. The experience in the United States over the past three decades – and rapidly being replicated elsewhere – has been a rise in individual body weight due to declining physical activity and excess food intake. The prevalence of obesity, where body mass index (BMI) >30 is a generally accepted indicator, has increased significantly such that, according to the US Centers for Disease Control and Prevention, half of the adults in the USA have a BMI ≥ 30.

Considering that each kilogram of body fat contains 7778 kcal of stored energy, and making a conservative assumption that each person in the USA carries 4.5 kg of extra fat, as a nation the USA has stored at least 9.9 trillion kcal as body fat, representing unnecessary and potentially destructive use of our agricultural and fuel resources, and a huge reservoir to be released as CO_2 at our death.

Drawing on available data, the study estimated that 18 per cent of available food in the contemporary US system comprised luxus consumption, representing an ecological footprint of 0.36 ha of farm and ocean land per person. This represents an area used for overconsumption and waste of food in the USA that is greater per capita than the entire footprint of Bangladesh. Using carbonated beverages as one example, Blair and Sobal calculate that US consumption of 75.7 litres beyond need, representing the increase in per capita availability since 1983, required huge amounts of energy, fertiliser and pesticides to produce the 1.97 million tonnes of corn (maize) used to make the high-fructose corn syrup (HFCS) that is employed as a sweetener. There is also the huge amount of energy needed for bottling soft drinks in plastic.

> "The levels of luxus consumption are much larger than many would expect, and have a potent impact upon land use, soil loss, energy expenditure, pollution, body fat, and degenerative diseases in the US."

Source: adapted from Blair and Sobal (2006).

about environmental degradation and obesity continue to rise, we need more detailed research on patterns of overconsumption and other forms of waste as critically important insights into the contemporary global food system.

6. Summary

This chapter explores those stages beyond the farm gate associated with the manufacture of final foods, their distribution, retail and disposal, and highlights many of the environmental dimensions of these activities. It began by considering how social, economic and technological changes proceeded in a co-evolutionary manner, such that a widening array of time-saving and convenient food products became more widely available and helped to shape many changes within the domestic arena. The significant commercial growth of the food-manufacturing industry in the second half of the twentieth century was a consequence of its considerable broadening of the range of food products and its international expansion into new markets. Then, with the rapid development of supermarket retailing from the 1960s, the food system became even more tightly integrated.

Focusing upon refrigeration and packaging, this chapter highlights the ways in which technological innovations create certain path dependencies. Thus the cool chain has become a major component of the global food system, with fresh fruit and vegetables being chilled within moments of picking and remaining in sub-ambient temperatures until their preparation for eating, some days or weeks later, usually thousands of miles away from where they were grown. So much of our food is now locked into chill-chain distribution logistics that we have become highly refrigeration-dependent, with a significant, and to some extent avoidable, environmental cost. Packaging, too, has drawn us into less than optimal resource use, with the beverage industry in particular benefiting from sales of individual servings of water.

This chapter introduces the notion of LCA and provides a couple of examples of everyday final foods to illustrate the different environmental impacts that might be attributed to individual ingredients and industrial processes. It then explores movements of food, especially those associated with the retail sector. The continuous search for improved efficiencies and cost reduction has led retailers to engage in significant technological innovation as regards stock control and transport logistics. Nevertheless, such environmental improvements as are achieved, for example with regard to processing technologies, are generally outweighed by the rising volumes of goods being moved over greater distances. The chapter also reveals that there are a number of more complex considerations than being able to reduce environmental impact to a singular measure such as food miles, not least because much depends upon the shopping behaviour of consumers.

The chapter closes with a discussion of food waste that is clearly identified as occurring at each stage along the food supply chain. However, the most morally problematic and largest sources of waste appear to be through retailer outgrades of fresh produce, discards arising from convenience foods, and poor stock management and overeating by consumers. This excess consumption and waste of food should be kept in mind as chapter six explores the issue of food security in its various forms.

Further reading

Pollan, M. (2006) *The Omnivore's Dilemma: The Search for a Perfect Meal in a Fast-food World.* London: Bloomsbury.
Michael Pollan is simply a great writer and is a pleasure to read, even if one might take issue with aspects of his gourmandising.

Schlosser, E. (2001) *Fast Food Nation: What the All-American Meal is Doing to the World.* London: Penguin.
One of the first serious exposés of the food industry, and remains a must-read.

Stuart, T. (2009) *Waste: Uncovering the Global Food Scandal.* London: Penguin.
Although there are plenty of autobiographical "tales from the dumpster", Stuart has done an excellent job in pulling together so much detail, and the case he makes requires a serious response by industry and public policy.

Food & Water Watch (2007) *Take Back the Tap: Why Choosing Tap Water over Bottled Water is Better for Your Health, Your Pocketbook, and the Environment.* Washington, DC: Food & Water Watch, www.foodandwaterwatch.org/water/bottled/pubs/reports/take-back-the-tap

And on the web

Polaris Institute, Inside the Bottle – The People's Campaign on the Bottled Water Industry: www.insidethebottle.org

6 Rethinking food security

Contents

1. **Introduction**
2. **Tracing the evolution of food security**
   Food poverty
3. **Feeding the world: the challenge of population growth**
4. **The interlocking of food and energy security**
5. **Climate change, vulnerability and food security**
6. **Globalisation and food security**
   Production
   Consumption
7. **Food sovereignty**
8. **Summary: rethinking food security?**
Further reading

Boxes

6.1 Understanding people's entitlements to food

6.2 Food poverty
6.3 Economic growth and dietary change
6.4 The world's growing food-price crisis
6.5 Impacts of increased fertiliser prices in Malawi
6.6 Higher temperatures in the Sahel
6.7 The case of coffee
6.8 The downside of export horticulture
6.9 Production problems in achieving export quality
6.10 Foreign direct investment in the manufacture and retail of processed foods
6.11 Rethinking agricultural knowledge, science and technology

Figures

6.1 World grain production per person, 1950–2007
6.2 World grain stocks as days of consumption, 1960–2007

1. Introduction

More than 920 million people in the developing world do not have enough to eat, and a further 34 million people in the industrialised countries and economies in transition also suffer from chronic food insecurity. Food insecurity is generally taken to mean a dietary intake of insufficient and appropriate food to meet the needs of growth, activity and the maintenance of good health. In addition to those suffering from chronic hunger, many millions more experience food insecurity on a seasonal or transitory basis. Prolonged periods of insufficient food intake result in protein-energy malnutrition with loss of body weight, reduced capacity to work and susceptibility to infectious, nutrient-depleting illnesses such as gastro-intestinal infections, measles and malaria. Even mild undernourishment in children can lead to delayed or permanently stunted growth. There are almost 200 million children in the world displaying low height-for-age, with almost half of the children of South Asia failing to reach the weights and heights considered to represent healthy growth (FAO 2006, 2008).

In a context where the world produces enough food for all, why has it proven so difficult to reduce the number of hungry and malnourished people in the world? And why, given the undertakings that were made at the 1996 World Food Summit to cut the number of malnourished people (then 840 million) by half by 2015, does that objective look increasingly unrealistic? Moreover, since that Summit, the number of overweight and obese people has rapidly overtaken the number of hungry, with the greatest proportion in developing countries (Hawkes 2006). It might be argued that *mal*-nourishment, meaning badly nourished, concerns both the underfed and overfed, and raises profound questions about health, wellbeing and food security across the nutritional spectrum.

The purpose of this chapter is to explore what we mean by food security, and to understand how it is intricately intertwined with a range of other pressing concerns such as existing and projected levels of population and consumption, as well as some of the global challenges that we have discussed previously, energy security and climate change amongst them. It will become apparent that economic uncertainties and increasingly complex, turbulent and unpredictable environmental futures not only make the goal of strengthening food security ever more vital, but highlight the need for fresh and critical thinking in ensuring that all people, but especially the poorest, gain greater control in meeting their food needs. Ultimately, how we feed ourselves in the years to come will require a broader and more robust conceptualisation of food security than we have had hitherto. The notion of food sovereignty may make a valuable contribution to this thinking.

2. Tracing the evolution of food security

It has been suggested that there are approximately 200 definitions and 450 indicators of food security (Hoddinott 1999), and this diversification of meaning reflects its wide interest as an object of study across a broad spectrum of academic disciplines (including the social, agricultural and nutrition sciences) and its application as a policy tool in various sectors of government. In some ways, the diversification and evolution of "food security" mirrors the growth and proliferation of meanings attached to the term "sustainable development". In both cases, it is as well to look closely at the interests and motivations of those using the terms rather than taking them at face value.

Although it had been largely confined to use by scholars and practitioners of development (concerning social, economic and cultural change in countries of the South), more recently food security has found its way into policy circles and documents concerned with food systems in countries of the North. Rising oil and food commodity prices have caused many countries that have long considered themselves highly food secure to take stock of their reliance upon global supply chains that deliver a high proportion of their food needs. In the UK, food security has proven a focus of attention not only within the Department for Environment, Food and Rural Affairs (Defra 2008c) and for the Royal Institute of International Affairs (Ambler-Edwards *et al.* 2009; Evans 2009), but also for the UK government (Cabinet Office 2008). We examine the particular fears and vulnerabilities of the North a little further on; first it is necessary to trace the changing meaning of the term.

Food security first appeared at the 1974 World Food Conference, where it was defined as: "availability at all times of adequate world food supplies of basic foodstuffs . . . to sustain a steady expansion of food consumption . . . and to offset fluctuations in production and prices" (Clay 1997: 9). The definition reflects the circumstances of the early to mid-1970s, when drought across many major grain-producing regions of the world led to heavy demand on international grain markets. Famine stalked the Horn of Africa and the Sahel, as well as South Asia, and encouraged the popular view that food insecurity was both demographically induced ("overpopulation") and environmentally determined (caused by drought, flood or soil erosion). The unfolding humanitarian disasters of the 1970s and 1980s, in which more than 2 million people died, did stimulate detailed analyses of the intersection of hunger, poverty, conflict, environmental degradation and the coping strategies of those affected. While detailed local-level studies revealed the limitations of overly deterministic causal relationships, they recognised that problematic long-term trends might combine with "trigger" events (e.g. drought, armed conflict or economic crisis) to tip already stretched local societies into acute distress. Thus a local society vulnerable to seasonal food insecurity, marked by a hungry period before the next harvest, might be tipped into a situation of structural malnutrition and chronic food insecurity by such an event. Understanding the circumstances experienced by the most vulnerable was a particular feature of the analysis of Amartya Sen (Box 6.1).

Box 6.1

Understanding people's entitlement to food

In 1981, the distinguished, Nobel Prize-winning economist Amartya Sen published his book *Poverty and Famines*, which forcefully demonstrated that hunger and starvation are not an inevitable consequence of a decline in the *availability* of food, but rather reflect the circumstances of people not being able to secure *access* to food. This can be explained, argues Sen, by understanding people's entitlement relations. On the basis of their initial endowments in land, other assets and labour power, a person has entitlements to his or her own production, the sale of labour for wages or the exchange of products for other goods (such as food). Under "normal" conditions these entitlements provide the basis for survival. But new circumstances, such as the occurrence of drought, may have an unfavourable impact on them. In that case, with the prolonged failure of rains and in the absence of irrigation, field crops simply shrivel and die. For local people who ordinarily earn wages by working in those fields, and whose labour is no longer needed, at least until the return of the rains, their main entitlement to food (their wages) collapses and they become highly vulnerable to hunger. A similar predicament confronts those with a few livestock. In the absence of adequate grazing, animals weaken and their value drops. Meanwhile, under the law of supply and demand (exacerbated by the opportunism of intermediaries), grain prices soar and the exchange rate of grain for animals deteriorates rapidly. This is a situation faced by all who must purchase their food needs and who experience a collapse in their entitlement relations.

Sources: adapted from Sen (1981); Drèze and Sen (1989).

By the mid-1980s, food security was no longer limited to the arithmetic of food supply and population, but became associated with efforts to understand the material basis of people's survival strategies. Agronomic research analysed the strengths of local farming systems and investigated the environmental and technical knowledge of farmers. Meanwhile, social anthropologists examined the role of non-agricultural household activities and used the term "livelihood strategies" to describe the different kinds of work that collectively contributed to securing people's access to food.

Thus the 1980s witnessed a growing interest in household-level food security, using livelihood and gender analysis to understand how vulnerable individuals and households cope with environmental, economic and political uncertainty, whether chronic or on seasonal, periodic or irregular time scales. Recognising the influence of external factors, such as economic shocks, on local food provisioning systems underlined the importance of appreciating the interconnections between the individual, local, regional, national and international levels. This has effectively meant that food security analysis can be applied at a variety of scales from the global to the household level, with each requiring its appropriate policy measures to ensure it is achieved.

Initially, food security was concerned with basic foodstuffs, principally high-calorie staples such as cereals and tubers, to resolve problems of protein-energy malnutrition. By the late 1980s, however, health and nutrition research had highlighted that nutritional wellbeing could not be assured from calorie consumption alone. A better understanding of human physiological capacity to utilise food intake revealed the crucial role played by disease, especially gastrointestinal infections, which impair the body's ability to absorb both micro-nutrients and calories. Moreover, nutrition research revealed the importance of micro-nutrients to human wellbeing. Iron deficiency, for example, is associated with a lack of physical energy and difficulties in concentration. While anaemia is a common condition requiring attention during pregnancy, it is estimated to affect 70 per cent of non-pregnant women in India, 50 per cent in Sub-Saharan Africa, and up to 2 billion people worldwide. Iodine deficiency is thought to be responsible for an estimated 655 million cases of goitre and 6 million people suffering from cretinism worldwide. A deficiency in vitamin A is believed to affect one-quarter of all children under five in developing countries, being a major cause of childhood blindness, and responsible for 1 million child deaths per year (DeRose and Millman 1998).

By the time of the World Food Summit, held in Rome in November 1996, the definition of food security had evolved further to reflect social and cultural influences over food preferences:

> Food security, at the individual, household, national, regional and global levels is achieved when all people, at all times, have physical and economic access to sufficient, safe and nutritious food to meet their dietary needs and food preferences for an active and healthy life.
>
> (FAO 1997)

The Rome Declaration, which the Heads of State agreed at the Summit, pledged political will and common commitment to achieve food security for all, with an immediate target of "reducing the number of undernourished people to half their present level no later than 2015" (FAO 1997: 83). Yet, as we have seen, rather than moving toward the target of 400 million people, the ranks of the hungry have swollen by at least 100 million since 2007.

This is not from lack of hand-wringing, as food security has increasingly come to be seen as part of a wider concern not just for human welfare, but as a basic human right. In Objective 7.4 of the World Food Summit Plan of Action, a call was made for the implementation of Article 11 of the 1967 International Covenant on Economic, Social and Cultural Rights, which affirms "the right of everyone to an adequate standard of living for himself and his family, including adequate food, clothing and housing". Clause E of the Objective calls upon the UN High Commissioner for Human Rights "to better define the rights related to food in Article 11 of the Covenant and to propose ways to implement and realize these rights . . . taking into account the possibility of formulating voluntary guidelines for food security for all" (FAO 1997: 122–23). After several years of high-level consultations and meetings, the FAO Council adopted by consensus the "Voluntary Guidelines to Support the Progressive Realization of the Right to Adequate Food in the Context of National Food Security". A specialist Right to Food unit now exists within the FAO. Yet more than 920 million people remain undernourished.

This demonstrates a fundamental problem with food security: that despite the efforts to enshrine the human right to adequate food, there is no effective mechanism to ensure its fulfilment. International human rights instruments are concerned primarily with the responsibilities of states to their own people, not to people elsewhere. The principle of national sovereignty, which underpins international law, generally restricts the intervention of foreign governments even when states may be failing to provide for and to protect their own citizens. Consequently, food security persists largely because of a failure of government at national level and a lack of international political will. This suggests that, despite ongoing efforts to establish a legal right to food within international law, ultimately more immediate and practical solutions for strengthening food security are more likely to be found at local level. This is an issue to which we return later.

Food poverty

Despite an apparent focus on the least developed countries, it is instructive to note that food insecurity is far from unknown in the most developed countries of the world, where millions of people are undernourished and food insecure. This further underlines that hunger is everywhere a function of social inequality, poverty and the failure of entitlements. In developed countries with established welfare services, social policy has long recognised the problem of food *poverty*, rather than food insecurity. Here, people's relationship to food is conceivably more complex still: on low wages or welfare benefits, people lack sufficient money to buy appropriate food; yet they are surrounded by the thousands of products of the modern food system. Moreover, many of their fellow citizens are striving to reduce, rather than increase, their calorie intake. Some illustrations of food poverty can be found in Box 6.2.

In this respect, food poverty can be considered a measure of both absolute and relative social deprivation. Absolute poverty means that people do not have enough money to pay rent, heat their living space ("fuel poverty"), buy clothes, afford transport and generally look after themselves, including buying sufficient food. Relative deprivation refers to circumstances where people lack the resources needed to enjoy the living conditions and amenities, and to access the types of diet, that are customary, in the society to which they belong. Effectively, they are excluded from prevailing patterns of social life (Riches 1997).

For example, people may not have enough money to *acquire* sufficient quantity or adequate quality of food, given other pressing demands on the household budget. Besides the items mentioned earlier, meeting the needs and wants of children, who are so heavily influenced by peer group and society, poses a particular challenge for parents on low income. Cycles of indebtedness, especially where people have taken

---

Box 6.2

Food poverty

Hunger and food insecurity are prevalent in the United States. The US Department of Agriculture reports that, in 2005, 11 per cent of all US households were "food insecure" because of a lack of sufficient food. Black and Hispanic households experienced food insecurity at far higher rates (22.4 and 17.9 per cent, respectively) than the national average.

In 2003–04, requests for emergency food assistance increased by about 14 per cent in the 27 cities surveyed by the US Conference of Mayors. About 20 per cent of the demand for food went unmet. Fifty-six per cent of those requesting assistance represented families with children; 34 per cent of adults requesting assistance were employed.

In 2003, 21.2 million individuals participated in the US Food Stamp Program; however, this represented only 60 per cent of people eligible to receive food stamp benefits. The average monthly food stamp benefit was US$83.77 per person.

Research suggests lower rates of obesity and overweight in neighbourhoods where supermarkets offering healthier food choices are present. This access is not even, however: low-income and minority areas contain fewer supermarkets on average; these areas also tend to have a higher density of convenience stores offering fewer healthy choices and higher prices, and fast-food outlets. Because these communities experience lower vehicle-ownership rates, problems of access are exacerbated.

Source: adapted from APA (2007).

loans from local moneylenders at usurious rates of interest, have enormous repercussions for household wellbeing. With food usually the most "flexible" item in the household budget, the residual funds available after essential bills and debts have been paid can then be spent on food. It is little wonder that shopping for food may then reflect a desire to maximise calories and the satisfaction of "feeling filled" for every pound, euro or dollar spent. Such "cheap" food may be more *affordable*, but is often the least healthy and may be a major determinant in obesity, high blood pressure and the onset of diabetes.

Secondly, people may lack *access* to shops selling food at reasonable prices where, usually, the large chain supermarkets offer the most competitive prices for everyday items. Yet food policy analysts in the UK, for example, have highlighted the difficulties faced by low-income urban communities marked by limited mobility in getting access to fresh, healthy food. Confronted by the policy of supermarkets to relocate to edge-of-town sites requiring access to a car, the term "food desert" has been used to describe those urban areas of social exclusion deprived of food retail outlets that can offer a wide range of fresh, healthy produce (Lang and Caraher 1998).

A third aspect of food poverty concerns the *ability* of people to make appropriate purchasing choices and then to prepare that food in socially acceptable ways to deliver nourishment. For example, being trapped in a long-standing situation of food poverty understandably engenders a sense of disempowerment that, in turn, gives rise to a lack of interest in cooking and results in above-average consumption of ready-made food and snacks. Given that much of the urban landscape in the world today is dominated by symbols and signs for fast food and carbonated beverages, and that such products are utterly ubiquitous and offer instant gratification, it is unsurprising that food poverty has also become linked to the issue of obesity. This is not to equate them: the consumption of high energy-density "pseudo" foods is certainly cheaper than healthy food, and while a diet largely comprising such items may give rise to health problems, overweight and obesity is not inevitable. This largely depends upon levels of physical activity, and this reflects the wider everyday geographies of people's lives. In this regard, living in environments that do not facilitate physical activity (access to outdoor recreation, green space, sense of security in the community) is as important to the risk of obesity as unhealthy food consumption practices (Poortinga 2006).

While the existence of food poverty in wealthy, highly developed countries testifies to the failure of welfare policy and even to effective, socially inclusive national governance, its solution requires more than enhanced handouts. This is why food security has to be approached as an issue of social justice as well as a matter of human rights. In the United States, the Community Food Security Coalition promotes an empowerment-driven and justice-oriented approach to food security, which it believes is best operated at the community level. Its definition is that community food security is a condition in which "all community residents obtain a safe, culturally-acceptable, nutritionally-adequate diet through a sustainable food system that maximizes community self-reliance, social justice, and democratic decision-making" (Winne n.d.: 2).

In summarising this section, it is apparent that the meaning of food security has evolved: from circumstances where an aggregate supply of calories at national or regional level was once sufficient guarantee that hunger was eliminated, to a situation deeply entwined with human rights and the struggle of communities to define their own particular food needs. In the sections that follow, this chapter explores a number of significant challenges for achieving global food security in the decades ahead, and demonstrates the value of developing a broad-based and integrated approach.

3. Feeding the world: the challenge of population growth

One significant approach to the study of food security over the past two or three decades has derived from an *a priori* concern with demographic growth and the capacity of the Earth to provide sufficient food to support the global human population. This approach is generally labelled "neo-Malthusianism', tracing its intellectual lineage back to the English parson Thomas Malthus who, in 1798, published *An Essay on the Principle of Population*. This "First Essay" is known largely for its rather stark postulation that human populations tend to grow in a geometrical ratio (1, 2, 4, 8 . . .) while food supply increases in an arithmetical ratio (1, 2, 3, 4 . . .). Inevitably, the balance between them can be maintained only by upward pressure on the death rate, brought about primarily by hunger, starvation and disease. As Dyson (1996) explains, Malthus' generalisation of a linear, arithmetic increase in food production was entirely reasonable given his implicit assumption that the supply of cropland was fixed, and that this increase represented a steady improvement in agricultural yields.

It has been a particular concern of neo-Malthusians that growth in cereal yields slowed significantly from the 1980s. This followed the upturn in yields attributed to the Green Revolution's technological package of high-yielding varieties of seed, chemical fertilisers and pesticides, and irrigation that took off in the 1960s and 1970s. Neo-Malthusian writers, such as Lester Brown, draw particular attention to a critical turning point during the mid-1980s, when the world's population began growing more quickly than production of cereal grains. This is shown in Figure 6.1 as a gradual decline in the global output of grain per person. Now, one might read into such trend lines all sorts of catastrophic scenarios in which the global rate of population growth is tipping us all toward a future of hunger, food scarcity and insecurity. On the other hand, it might be more useful to move beyond the rather clumsy global "population–food" equation in order to understand better the deeper underlying trends.

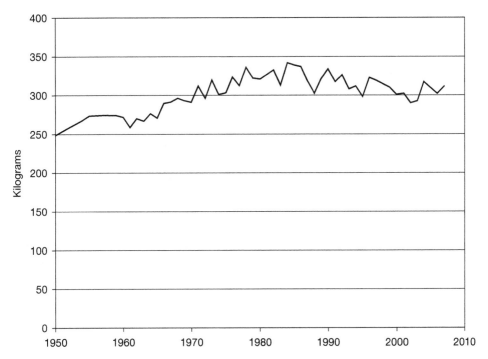

Figure 6.1 *World grain production per person, 1950–2007*
Source: Earth Policy Institute, www.earth-policy.org/data_center.

First, world grain production per person is the result of combining harvested area of grains per person and an average level of yield (output per unit area of land). Data show that yield has generally climbed steadily, while the area harvested per capita has fallen – 90 per cent of the increase in grain output since the 1950s can be attributed to yield, with only 10 per cent due to an increase in the area harvested. Yet it is not simply the case that population growth has been inexorably outstripping the rate of expansion of agricultural land. Beginning in the 1980s, for example, the EU paid farmers to set aside land from cereal production in order to address structural overproduction, the high costs of maintaining large food stocks ("grain mountains") and the prevailing low market prices of cereals, a policy that had its counterpart under successive Farm Bills in the USA. Consequently, "retiring" land from cereal production has served to offset expansions elsewhere and contributed to the modest expansion of the total harvested area.

Second, cereal gains, though important, are not the only food consumed by the global population, and demand for them may fall as incomes rise and globalisation presents new food products encouraging dietary diversification. Per capita demand for rice has fallen in East and South Asia, for example, and given the weight of these regions in world rice consumption, accounts for the flattening of the global trend line for that cereal (Box 6.3). Wheat, however, continues to rise in terms of per capita demand as not only traditional wheat-consuming societies grow in population, but also new countries with limited production capacity incorporate bread and pasta into their diets. But, above all, it is livestock products – principally meat, but also milk and dairy – that have witnessed the biggest expansion. In the South, meat consumption has been growing at over 5 per cent per year over the past few decades, with milk and dairy growing at 3.5–4 per cent per year according to FAO (2006). Oil crops have also grown at over 4 per cent per year during the past twenty years, for both food and animal feedstuffs (soybean meal). This sector has had a particular role to play in the expansion of agricultural land, accounting for over half the total expansion of the cultivated area under all crops over the period 1979/81 to 1999/01 (FAO 2006). Consequently, to focus exclusively upon cereals is to overlook

---

Box 6.3

Economic growth and dietary change

Higher incomes, urbanisation and changing preferences are raising domestic consumer demand for high-value products in developing countries. The composition of food budgets is shifting from the consumption of grains and other staple crops to vegetables, fruit, meat, dairy and fish. The demand for ready-to-cook and ready-to-eat foods is also rising, particularly in urban areas. Consumers in Asia, especially in the cities, are also being exposed to nontraditional foods. Due to diet globalisation, the consumption of wheat and wheat-based products, temperate-zone vegetables and dairy products in Asia has increased.

Today's shifting patterns of consumption are expected to be reinforced in the future. With an income growth of 5.5 per cent per year in South Asia, annual per capita consumption of rice in the region is projected to decline from its 2000 level by 4 per cent by 2025. At the same time, consumption of milk and vegetables is projected to increase by 70 per cent and consumption of meat, eggs and fish is projected to increase by 100 per cent.

Source: adapted from von Braun (2007), 1.

important developments in other commodity sectors and, when taken as a whole, food production has continued to keep pace with population growth.

There is a third and positive issue that should also be noted. That is, that while global population did indeed grow swiftly during the 1960s–1980s, it is now beginning to slow. As a prospective study for the FAO puts it, "longer term projections suggest that the end of world population growth may be within sight by the middle of this century" when world population may reach 8.9 billion.

It is instructive, nevertheless, to note that at the time Malthus published his first essay, the world's population was at 1 billion, and it took 130 years for it to double. The development and widespread dissemination of vaccines, antibiotics and other public health measures during the middle of the twentieth century enabled a third billion to be added in thirty years, and a fourth in just fifteen years. Demographic momentum added a fifth and a sixth each within twelve years, and world population is likely to reach 7 billion by 2010. The prospect that it will plateau at around 9 billion by 2050 offers the FAO to reflect that "over the next 50 years world agriculture may be transiting to a future when global population growth will no longer be the major driving force for further growth in world food demand and production. This has consequences for the rate at which further pressures on land and water resources and the wider environment will be building up" (FAO 2006: 1).

Yet, while we may take some encouragement from evidence of a slow down in global population growth and the prospect of relieving some of the pressure on the global resource base used to produce food, in reality this disguises considerable regional differences around the world. For example, many of the most developed countries and economies in transition are experiencing a decline in fertility rates to below replacement levels. Yet, while Japan, Russia and Germany are projected to decline by 25, 22 and 13 per cent, respectively, over the period 2008–50, the populations of Uganda, Burundi and Niger will more than triple in that time (Population Reference Bureau 2008). In this regard, we can think of a demographic bifurcation of the world between those countries where population numbers are stabilising, and those where conditions of poverty, uncertainty and insecurity make it difficult for people to move beyond short-term survival. Taken as a whole, a focus on global population numbers does not tell us much about the prospects for food security in any particular part of the world. A preoccupation with numbers of people serves to disguise extraordinary differences in the levels of consumption within, as well as between, different countries.

For example, researchers at the International Food Policy Research Institute (IFPRI) have long drawn attention to the consequences of strong economic growth throughout the developing world, especially in Asia and particularly in China and India. In the former, a sustained period of rapid economic growth has made a significant impact in raising average household incomes. This has also been accompanied by a considerable shift of population from rural to urban areas in search of jobs in the new manufacturing sector. With more disposable income and access to a greater variety of foods, consumption practices have changed significantly in keeping with the long-established norm that as incomes rise, consumers do not eat more basic staples but shift toward higher protein sources such as meat and dairy products, vegetables and fruit, and prepared foods. In China, meat consumption per capita has doubled and milk consumption has increased by a factor of four over the period 1990–2006 (von Braun 2007). Given that a "livestock revolution" is under way throughout the South to meet this demand for meat and dairy products, and that more than one-third of world cereal production is used to feed such animals, it is changing consumption practices that will have a greater influence on agricultural resources than human numbers *per se*. Finally, this rising demand for livestock feed has undoubtedly contributed to the recent rises in global food prices, as we will see later.

4. The interlocking of food and energy security

During 2007 and 2008, food security concerns were heightened as a result of a dramatic increase in agricultural commodity prices and the sharp decline in world grain stocks as measured by the number of days of consumption in store (Figure 6.2). From the beginning of 2003 until the middle of 2008, maize and wheat doubled in price and rice tripled in price. Ordinarily, neo-Malthusians would be quick to use this as evidence to support their case of a food–population crunch. Yet, although there were other contributory food factors (demand for meat and therefore livestock feed), the episode most clearly demonstrated, as never before, the interlocking nature of food and energy markets, and revealed that there are yet more powerful claims on global food stocks than the demands of hungry people.

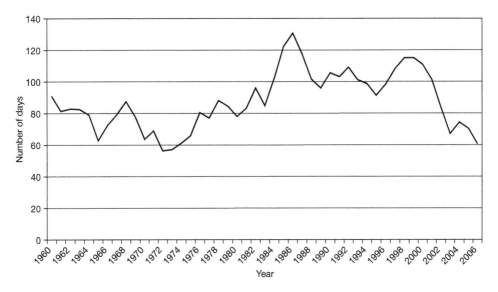

Year

Figure 6.2 *World grain stocks as days of consumption, 1960–2007*
Source: Earth Policy Institute, www.earth-policy.org/data_center.

Box 6.4

The world's growing food-price crisis

Rocketing food prices – some of which have more than doubled in two years – have sparked riots in numerous countries recently. Millions are reeling from sticker shock and governments are scrambling to staunch a fast-moving crisis before it spins out of control. From Mexico to Pakistan, protests have turned violent. Rioters tore through three cities in the West African nation of Burkina Faso last month, burning government buildings and looting stores. Days later in Cameroon, a taxi drivers' strike over fuel prices mutated into a massive protest about food prices, leaving around 20 people dead. Similar protests exploded in Senegal and Mauritania late last year.

Source: adapted from Walt (2008).

In July 2008, the price of a barrel of Brent crude reached an all-time high of $147. While commentators look for singular triggers to explain such price hikes – problems in the Nigerian Delta as rebels interrupt supplies, for example – the more fundamental fact is that rising levels of oil consumption have effectively pushed up against the prevailing ceiling of production. With the model of insatiable demand for fuel to move increasing quantities of goods ever-greater distances now firmly implanted in China and India as well as in the traditional heartlands of energy profligacy, yet with global oil output stuck at around 84 million barrels per day, market prices reflected this excess demand over supply. This inability to raise oil output suggested not only that the moment of "peak oil" had arrived (see chapter four), but also that the ensuing financial crisis that began in September 2008 was partly triggered by speculation that ultimately rested upon an infinite supply of cheap energy. Unfortunately, however, the rise in the cost of oil (as well as natural gas, which tracks oil fairly precisely) passed through into the price of food very quickly. The continuing instability in global food prices appears to be reinforcing the view that food and energy markets are now closely intertwined.

Identifying the particular, individual drivers that served to translate the high cost of oil into the rising price of food became the focus of a considerable amount of policy analysis during 2008 and 2009. While a number of contributory factors were identified – adverse weather conditions affecting grain harvests in Australia, the USA and Central Asia; the activities of financial speculators in buying agricultural commodity futures; dietary change with increased demand for animal feeds; and low global grain stocks – the single largest role was assigned to biofuels. Even a World Bank document suggested that no less than three-quarters of the 140 per cent increase in the Bank's food prices index from 2002 to 2008 was caused by biofuels and related effects (Evans 2009). As discussed in chapter four, increased demand for agricultural commodities for biofuel production competes with the existing food use of those commodities. In the case of sugar cane in Brazil, which has been used for the production of ethanol since 1970 and which is not strictly a food staple, its price closely reflects the price of ethanol. But with growing demand for corn (maize) by ethanol distilleries in the USA, price rises translate directly into the costs of food.

Evidence of the interlocking nature of the global markets for energy and food is reflected in the way high energy-price fluctuations are mirrored by movements in the price of food. In the past five years, according to von Braun (2007), price variations in oil seeds, wheat and corn have doubled the levels of fluctuations seen in previous decades.

When oil prices range between US$60 and $70 a barrel, biofuels become competitive under existing technologies, pretty much everywhere; as oil rises further, biofuels become an ever more attractive alternative. Given that feedstock (sugar cane, corn, palm oil, etc.) represents between 50 and 80 per cent of overall costs of biofuel, depending on whether it is ethanol or biodiesel, increases in feedstock costs have a direct impact on the comparative advantage and competitiveness of biofuels. In other words, biofuels contribute to feedstock price changes, but are also vulnerable to such changes. All of this suggests a much more complex future, in which there will be growing competition between agricultural production for food or fuel, the relative parameters being set by the prevailing price of oil and the pace of scientific developments in rolling out second- and third-generation biofuel technologies.

In the meantime, food prices have risen significantly. In the UK, food price inflation reached a peak of 12.8 per cent in August 2008, though it was somewhat lower in other EU member states (e.g. Germany, 7.4 per cent) and the USA (7.1 per cent). This meant that a British family who spent £100 per week on food in 2007 had to spend £634 more during 2008 to consume the same basket of goods (Ambler-Edwards *et al.* 2009). But for people in low-income countries of the South, where many spend up to 70 per cent of their income on food, food price inflation was greater still, with the International Monetary Fund (IMF) calculating a rate of 67.5 per cent in Asia during 2007. Inevitably, rising prices were translated directly into increasing numbers of hungry and malnourished. With an estimated 854 million undernourished before the 2007–08 crisis, up to 105 million more people were affected by the rising cost of food up to 2007 (Evans 2009). Little wonder, then, about the civil unrest described in Box 6.4, events that are more rather than less likely to recur in the short to medium term, given the vulnerability of national food systems to wild upswings in price.

There are, of course, other dimensions of the interlocking nature of food and energy security besides the role played by biofuels. If there is one that must be addressed, it is the production of chemical fertilisers. As noted in chapter 2, almost half of the annual turnover of nitrogen in the world's crops is provided by synthetic nitrogen produced by the Haber–Bosch process using natural gas, and its importance has become somewhat irreplaceable. The manufacture of urea (the principal source of synthetic nitrogen) accounts for some 5 per cent of world natural gas production, which is used both as feedstock and process energy. As up to 90 per cent of the cost of the urea is determined by the price of natural gas, it follows that rising energy costs translate directly into higher fertiliser prices.

During 2007, the price of fertiliser rose at an even sharper rate than the cost of food, with urea, potash and phosphate prices rising by 1.5-, two- and threefold, respectively. Given that fertiliser costs account for the single largest item of expenditure in many industrialised agricultural systems in the North, farmers have been anxiously expecting to receive higher farm-gate prices in the hope that they are not to experience a further intensification of their cost–price squeeze. In the South, on the other hand, the situation is more acute still. Box 6.5 provides a detailed account of how higher fertiliser costs have a direct impact on food security in Malawi. Note how this rising import bill has repercussions for farmers, consumers and the Malawian government in its efforts to maintain domestic food production capacity.

Although this section could explore other aspects of this growing interlocking of food and energy security, the illustrations provided here demonstrate its significance and highlight the need for a much broader vision in addressing single issues such as rising fuel prices. Recall, for example, that one of the principal rationales by which the United States under President George W. Bush sought to expand domestic ethanol production was to reduce its external dependence on foreign oil. Although a laudable objective, the nitrogen-hungry corn crop does, in turn, increase dependence of the fertiliser industry on sources of natural gas, 20 per cent (and rising) of which is imported. Although there are other factors in play, it is incontestable that the rush to biofuels has contributed to rising global food prices and pushed more people into the ranks of the food insecure. Yet, if the demand for cheap fuel for automobiles has some responsibility not only for the failure to halve the number of malnourished, but also to increase their number, it may be that part of a solution to food security must be found in the field of energy. It might be argued that, given the imminence of peak oil, our insatiable demand for energy will have to be addressed if global food insecurity is not to spiral out of control.

5. Climate change, vulnerability and food security

Another truly significant challenge to improving future prospects for food security is presented by both systemic and cumulative aspects of global environmental change. Climate change and stratospheric ozone depletion are examples of systemic global change, while the depletion and contamination of freshwater resources, deforestation, desertification and desiccation of semi-arid environments, soil depletion and

Box 6.5

Impacts of increased fertiliser prices in Malawi

Average expenditure per person in Malawi is around US$250 per year. Many Malawians spend over 25 per cent of their income on their staple food, maize. 97 per cent of Malawian farmers grow maize, devoting around 50 per cent of their cultivated land to producing the crop, but 60 per cent of Malawian farmers consume more maize than they produce and have to supplement their own maize production with market purchases. Affordable maize prices are thus critical to the Malawian economy and to the wellbeing of the population. Malawian farmers face particularly high fertiliser prices due to the costs of transporting low volumes from the coast and then into rural areas – and as a result, farm-gate fertiliser prices can be more than double international prices.

Undertaking an evaluation of the prevailing prices during the 2006/07 and 2007/08 growing seasons, the authors reveal some of the complex repercussions associated with the rising cost of imported fertilisers. For example, post-harvest market prices in the past two growing seasons have fallen below the break-even price of maize, meaning that farmers do not recoup their full costs of production. Moreover, in 2008/09 the break-even price for profitable fertiliser use was expected to be more than double the break-even price of the previous year and, with an average nitrogen response rate, would be roughly equal to the expected price of importing maize from South Africa. Malawi was consequently facing the prospect of maize prices that would cause severe hardship and increase already high malnutrition and poverty rates.

The increase in international fertiliser prices also has a major impact on the country's ability to afford them. From 2003/04 to 2006/07, the cost of one 50 kg bag of urea without any subsidy cost a little over 10 per cent of median annual per capita rural expenditure (though few households had either savings or access to credit that would allow them to purchase fertiliser). The 2005/06 and 2006/07 subsidies led to increased production by making fertiliser purchase affordable. At 2008/09 prices, the same bag of urea, with no subsidy, would cost around 30 per cent of median per capital expenditure. To deliver the same volume of fertiliser at the same (affordable) subsidised price in 2008/09 as in 2006/07 would require government fertiliser subsidy expenditure of over US$200 million, over three times the cost in 2006/07, and around 17 per cent of the 2007/08 national budget.

Source: adapted from Dorward and Poulton (2008).

biodiversity losses are examples of discrete environmental changes that exert a cumulative effect at the global scale. The connections with food security are manifold, although the precise nature of interrelationships is complex and uncertain. For example, the concern of atmospheric scientists with the thinning of the stratospheric ozone layer is that this will be translated into increased penetration of ultraviolet (UV) radiation to the Earth's surface. It is understood that increased UV radiation can weaken the human immune system, leaving people more susceptible to infectious disease, while the effect on agricultural crops may be to lower yields. Fortunately, the Ozone Depletion Regime – comprising a significant package of measures that continue to be strengthened and extended, which are effectively eliminating ozone-depleting chemicals from production and use – is well established and is one of the truly successful examples of international environmental cooperation.

The challenges presented by climate change, however, are more far-reaching and intractable, and have not yet resulted in a comparable level of commitment needed by the international community to address the problem. For, as discussed in chapter four, the balance of evidence is now unequivocal that rising atmospheric concentrations of carbon dioxide and other greenhouse gases that derive from human activities, including food production, are having a discernible influence on the world's climate. Moreover, regional scenario-building exercises using general circulation and statistical crop models suggest that, without sufficient adaptation measures, many regions throughout the tropics are likely to experience a significant deterioration in food security. Remembering that more than 850 million people were undernourished prior to the recent food crisis, and that at least 1 billion remain vulnerable to food insecurity, the prospect of global warming and extreme weather events reducing the productivity of local agri-food systems is cause for acute concern.

Battisti and Naylor (2009) draw particular attention to the consequences of higher growing-season temperatures for agricultural productivity and food security. Making a link with the focus of the previous section, they remark that, while rapid growth in demand for food, animal feed and biofuels, coupled with disruptions in agricultural supplies caused by poor weather, created chaos in international markets, "the longer term challenge of avoiding a perpetual food crisis under conditions of global warming is far more serious" (*ibid*.: 240). Drawing on twenty-three global climate models that contributed to the Intergovernmental Panel on Climate Change Fourth Assessment Report (AR4) (IPCC 2007), their analysis shows that it is highly likely (> 90 per cent chance) that growing-season temperatures by the end of the twenty-first century will exceed even the most extreme seasonal temperatures recorded from 1900 to 2006 for most of the tropics and subtropics. Given that over 3 billion people live in these latitudes, many earning less than $2/day while depending upon agriculture, and with growing-season temperatures rising beyond historical bounds, the question that Battisti and Naylor pose is: can people in these regions ever hope to achieve food security?

Notwithstanding the attention given by general circulation models to increased threat of droughts, Battisti and Naylor highlight the consequences of sustained chronically high temperatures, that is, more than multi-day heatwaves, throughout the food-growing season. To help make their case, they draw upon the summer of 2003 in Western Europe, when an estimated 52,000 people died from heat stress between June and August, the majority in France and northern Italy, where mean summer temperature was 3.6°C above the long-term mean. With regard to food production:

> Record high day-time and night-time temperatures over most of the summer growing season reduced leaf and grain-filling development of key crops such as maize, fruit trees, and vineyards; accelerated crop ripening and maturity by 10 to 20 days; caused livestock to be stressed; and resulted in reduced soil moisture and increased water consumption in agriculture.
>
> (Battisti and Naylor 2009: 243)

With crop yields falling by more than 20 per cent (36 per cent in the case of maize in northern Italy), this example demonstrates the effect of high seasonal heat, albeit one without a marked deviation in seasonal rainfall. On the other hand, regions such as the Sahel, where drought and heat stress are tightly coupled, and where levels of food insecurity are already acute, the prospects of higher temperatures will exacerbate the vulnerability of the population, as Box 6.6 illustrates. Yet, with IPCC climate models projecting that growing-season temperatures by the end of the twenty-first century will exceed the hottest temperatures on record in each region, global food security will be severely jeopardised unless large adaptation measures are undertaken quickly. This means not only in the tropics and subtropics, but also across the North. As chapter four observes, in the mid- to high-latitude regions of the world, the medium-term rise in ground-level temperatures will lengthen the growing season, extend the northern limit of cereal cultivation, and may even offer opportunities to grow novel crops.

Yet, as warming proceeds, there will be increased evapotranspiration and soil moisture losses, excessively high temperatures and water shortages, and increased plant disease and pest outbreaks, all of which will threaten crop yields.

Box 6.6

Higher temperatures in the Sahel

Crop and livestock production play an essential role in the region's economy, employing roughly 60 per cent of the active population and contributing 40 per cent to gross national product. The Sahel suffered a prolonged drought from the late 1960s to the early 1990s that caused crop and livestock productivity to plummet, and that contributed to countless hunger-related deaths and unprecedented rates of migration from north to south, from rural to urban areas, and from landlocked to coastal countries. Although the Sahel's climate disaster was largely one of extended drought, the spectre of high and rising temperature lurks in the background.

Year-to-year temperature variability in the Sahel has been low during the past century, but the growing-season temperature has been very high, with long-term daily averaged summer temperature ranging from 25°C in the south to 35°C in the north. Moreover, temperatures have trended upward since 1980.

Maize yields have remained far below varietal potential, and millet and sorghum yields continue to stagnate. Hundreds of thousands of children and infants in the region still die each year from hunger-related causes, and malnutrition contributes to long-term mental and physical disabilities. Over recent decades, most of the region's poorest households have lost their livestock or other assets; they remain net consumers of food, and struggle to purchase staples even when they are available in the market. These households farm at an extreme disadvantage, irrespective of climate change.

Most worrisome for the Sahel is that average growing-season temperatures by the end of this century, and even earlier for some parts of the region, are expected to exceed the hottest seasons recorded during the past century. Such heat will compound food insecurity caused by variable rainfall in the region, and will increase the incidence of agricultural droughts. Even today, temperatures in the Sahel can be so high that the rain evaporates before it hits the ground.

New bounds of heat stress will make the region's population far more vulnerable to poverty and hunger-related deaths, and will probably drive many people out of agriculture altogether, thus expanding migrant and refugee populations.

Source: adapted from Battisti and Naylor (2009).

For those regions within the tropics and subtropics, however, climate change will bring unprecedented temperatures, effectively eliminating from production crops that may already be close to their limits for heat and water stress. Projections of changing precipitation patterns will affect rainfed agriculture, which comprises most of the cultivated land of Sub-Saharan Africa and the overwhelming part of Latin America. Climate change will also severely worsen the existing burden of disease and ill health, with the increased spread and transmission of vector-borne diseases such as malaria. As rural livelihoods collapse with the failure of crops and the death of livestock, migration to cities is likely to intensify, placing an even greater burden on public health infrastructures in urban areas, which are already seriously under-resourced and unable to deliver sanitation and clean water to existing populations. As Devereux and Edwards (2004: 24) note: "if the households and countries that stand to lose food production due to climate change are also those that depend most on agriculture and have fewest alternative sources of income, then falling harvests will certainly undermine household and national food security". In other words, the biggest losers from climate change are likely to be people who will be most exposed to the worst of its impacts and are the least able to cope.

Moreover, although food security is about *access* to food and not determined solely by self-sufficient production, projections suggest that the traditional food surplus-generating regions of North and South America and Western Europe will be unlikely to balance the food deficits of the tropics. Increased food aid is unlikely, therefore, to grow significantly from its existing levels – the World Food Programme of the United Nations currently struggles to feed fewer than 10 per cent of the world's malnourished – and the prospect of donor governments substantially increasing funding appears unlikely. However, food security will not simply be the result of a one-way outcome of biophysical changes in climate, but will reflect the social, economic, institutional and technological responses (and non-responses) to the challenge. Above all, it will be about reducing vulnerability and enhancing resilience, equitability and sustainability.

Hence the most urgent and vital challenge is the development of a twin-track approach combining mitigation and adaptation strategies. *Mitigation* concerns the implementation of measures designed to minimise greenhouse gas emissions while finding ways to enhance the sequestration and storage of carbon, measures briefly outlined in chapter four.

Adaptation, on the other hand, will be utterly central to building food security. The International Assessment of Agricultural Knowledge, Science and Technology for Development's Synthesis Report (IAASTD 2009) usefully distinguishes between two kinds of adaptation: autonomous and planned.

Autonomous adaptations are triggered by ecological, economic or welfare changes, and involves the implementation of existing knowledge and technology in response to the changes experienced. They are consequently largely extensions or intensifications of existing risk-management or production-enhancement activities, and include changes to cropping patterns such as adjustments to crop species and varieties; irrigation management and water conservation; and diversification of agricultural activities.

Planned adaptations, in contrast, represent a much more deliberate set of decisions recognising that change is under way and that action is required to maintain or achieve a desired state. At local level, they may be triggered by past experience of extreme events and the need to avoid a repeat of earlier difficulties. They represent a desire for increased adaptive capacity by improving or changing institutions and policies, possibly through investments in new technologies and infrastructure to enable effective adaptation activities. Specific policies aimed at reducing poverty and increasing livelihood security might involve the generation and dissemination of new knowledge, technologies and management practices tailored to anticipated changes in climate. Such policy-based adaptations will interact with other policy domains concerned with health, welfare, resources and governance, and may offer opportunities for mainstreaming climate change adaptation into policies intended to enhance broader societal resilience (*ibid.*).

Ultimately, it will be necessary for all countries, developed and developing, North and South, to review the ways in which their food systems might be vulnerable to climate change and to establish mitigation and adaptation measures to ensure the maintenance of food security. Climate-proofing the food system may pose some significant challenges, but one measure that would certainly contribute would be to increase domestic food production capacity in ways that might reduce not only greenhouse gas emissions, but dependence on extended supply chains from far-flung countries. Such a measure would, of course, raise questions about commitment to the globalisation project.

6. Globalisation and food security

"The world agrifood system is becoming increasingly globalised" claim von Braun and Diaz-Bonilla (2008: 1), citing as evidence increasing volumes of internationally traded food and a process of dietary convergence, amongst other things. Yet it is not at all clear that globalisation has strengthened food security, particularly for the poorest. While the brief discussion in chapter two highlights that globalisation is, strictly speaking, a multi-dimensional process involving social, cultural and technological changes, it is the economic realm, and especially international finance, that have been the key drivers of this project over the past three decades. In this section we explore the consequences of globalised production and consumption for food security in countries of the South.

Production

The model of comparative advantage, which argued that a country would be better off specialising in the production of things it was best at, rather than producing everything it needs, has been a constant of neoclassical economics and has been used to justify the continuing liberalisation of agri-food markets in the past three decades. This has entailed efforts to ensure that countries should have more interaction with the world economy, not less, and through this maximise production within their internationally competitive sectors. The generation of foreign exchange revenues through exports would then be used to purchase food needs through the global market. Chapter two outlines three ways in which agricultural production for global markets has developed in the South. First, this has been through the continuation of patterns of production established during the colonial era, when countries of the South supplied largely unprocessed tropical commodities such as sugar cane, cotton, beverage crops and bananas. A second, more recent development has been the movement into high-value fresh produce such as cut flowers, vegetables and aquacultural products; and a third is the emergence of large-scale agro-exports of bulk commodities capable of competing on international markets with the lowest-cost producers.

In all cases, we have seen how each of these different production strategies involves the diversion of land, water, labour and other resources into export crops rather than into the cultivation of domestic food staples. When tropical commodity prices were relatively buoyant and food imports were cheap during the 1950s and 1960s, this seemed a viable route to development for at least some developing countries. But, after the oil-price shocks of 1973 and 1979, there was a rapidly growing disparity in the terms of trade, that is, when comparing a typical basket of goods offered by countries of the South with that comprising manufactured and other essential items purchased from the North. This disparity grew throughout the 1980s and 1990s such that by 2002 the adjusted world market prices of all the major tropical agri-commodities were a fraction of their value in 1980. This pattern of "boom and bust" has been something of a distinguishing characteristic of the agricultural commodity market and is perfectly illustrated by the recent history of coffee, described in Box 6.7.

The case of coffee illustrates the dangers of structural overproduction arising from incentives to encourage new, "non-traditional" producers to participate in commodity markets, possibly on the basis of accepting lower returns. It also demonstrates the degree to which local production becomes shaped by vertically organised global supply chains, in which transnational buyers appear to exercise most of the decision-making power and derive most of the value. Although a commodity-chain approach has been criticised for being reductionist and deterministic, it does convey the disparity in the allocation of benefits from an internationally traded good, as well as the major disconnect that exists between the conditions within the realm of production and the world of consumption. At one end of the chain, falling prices result in reduced farm household income, which translates into less money to buy food (and medicine and school books) and poses a direct threat to the food security of coffee farmers. At the other end of the chain, coffee drinkers continue to pay the prices set by Starbucks (with more than 12,000 outlets serving 30 million customers per week in nearly forty countries) and Nestlé, which employs the charms of George Clooney to obscure the origin of the beans that comprise their drink or the circumstances in which they were produced. The retail price for coffee remained remarkably steady during the collapse of bean prices: little wonder that both of the above companies recorded substantially increased profits in 2001. This is why Fairtrade has become such an important development for many people, an issue discussed briefly in chapter seven.

For some countries, but by no means many, opportunities have arisen to diversify away from "traditional" agricultural commodities and into new, high-value agro-export production. Though guided by the same logic of

---

Box 6.7

The case of coffee

"After a frost destroyed much of the Brazilian coffee crop in the early 1990s, soaring coffee prices encouraged growers in countries like Kenya to expand rapidly. Within a few years, exports of Kenya's distinctive arabica beans were earning a quarter of a billion dollars a year. Unfortunately for Kenya, the same boom attracted other players, including Vietnam.

[While] Vietnam's climate is suitable only for the inferior robusta bean [it had] the backing of the big coffee companies, such as Nestlé, Procter and Gamble, Kraft, and Sara Lee, which together use almost 40 percent of the world's beans."

These companies had developed a process that stripped out some of the bad flavour of the robusta bean and used other added flavourings, such as vanilla, to mask the rest.

"Because robusta beans are 60 percent cheaper than arabica, the food industry now had [. . .] a new (highly profitable) food product made from a super-low-cost raw material."

Vietnam now became the model low-cost producer, making up in volume what it gave up on price.

"Money poured into the Vietnamese coffee industry": from the government of Vietnam, from the World Bank and from donor governments; Nestlé itself invested heavily. "The result of so much encouragement was predictable. Between 1990 and 2000, Vietnam coffee production soared from a few million tons to more than 16 million tons, overtaking Colombia as the world's number two producer and generating hundreds of millions of dollars in yearly export income."

With coffee production growing at twice the rate of demand, what was also predictable was the inevitable price collapse.

"Between 1997 and 2000, the composite robusta–arabica price fell from two dollars a pound to around 48 cents, well below many farmers' production costs."

For countries that depend on coffee for much of their export earnings – coffee provides Ethiopia with two-thirds, Uganda and Burundi with more than half of their export revenues – this was disastrous. More than half a million coffee labourers have lost their jobs worldwide; with unemployed farmers in Africa turning to poaching endangered animals for the bushmeat market; while in South America some have switched to coca, from which cocaine is made.

Source: adapted from Roberts (2008), 157–60.

international comparative advantage as a means to economic development, some countries have taken advantage of new supply chains delivering high-value produce into distant Northern markets. For example, Kenya, as noted earlier, has witnessed considerable expansion of horticultural produce, including green beans and cut flowers (Box 6.8); Chile has significantly expanded production of fruits (kiwis, grapes) and farmed salmon; and aquaculture output in South East Asia increased from less than 2 million tonnes (mt) in 1990 to more than 8 mt in 2006.

It is now well established that international trade in high-value produce illustrates the characteristics of a buyer-driven global commodity chain, with retailers utilising a network of subcontractors to perform all of the myriad specialised tasks that involve the growing, packing, transport and delivery of the produce to their stores. Corporate retailers seem to command great economic power while taking little direct financial risk.

Box 6.8

The downside of export horticulture

Outside Nairobi, thousands of acres of coffee trees have been dug up (since the price collapse) and replaced with massive open-air greenhouses filled with everything from peppers and baby corn to fresh flowers.

Kenya's horticultural sector is growing about three times as fast as the global food economy, and generated $700 million a year in 2007. And yet, like so much of the rest of the food economy, these trends have had a mixed effect on Kenyan food security.

It has been pulling land away from staple crops such as maize, while bidding up the price of inputs such as fertiliser. An increasing share of Kenyan farmland is devoted to products dedicated entirely for export, yet if the produce is rejected by European buyers, or suffers an outbreak of disease, the country will lose millions of dollars. Although rising jet fuel prices increased costs by 70 per cent in a single year, European supermarkets, locked in a price war for market share, refused to pay higher prices. As price pressure continues, more real cost reductions are borne by Kenyans themselves. This includes contracting small local farmers, using their produce to cushion the large commercial estate from any unexpected event; but it is the small farmer who is left with excess green beans when the supply quota is filled.

Sources: adapted from Roberts (2008), 165–6; Kenya Flower Council, www.kenyaflowercouncil.or

---

Their ability to respond quickly to perceived consumer concerns, whether food safety, labour practices or environmental impacts, has even seen retailers establish their own quality standards. The Euro-Retailer Produce Working Group Good Agricultural Practice (EUREPGAP) was established by a consortium of European retailers as a means to create a single, private standard for the production of fresh fruit and vegetables. Since 2008, the name GLOBALGAP has been used, indicating its global remit extending certification into a very wide range of agri-food exports. While as a standard it is primarily used to reassure consumers about the quality of their purchases, at the sites of production it imposes a wide and rigorous set of measures to which local producers must subscribe. It should be clear that imposition of "scientific" standards is not about creating development "partnerships" in which the expertise and empirically established indigenous knowledge of local farmers is recognised. Rather, it is about following the rules and practices laid down by the company in order "to satisfy the housewife" of the North (the gendered nature of this statement is explained in Box 6.9).

The bottom-line question that we need to ask, however, is: can involvement in such agri-export schemes enhance food security in the South, and if so, for whom? Is it legitimate to ask whether resources – land, water, labour and local knowledge – that might otherwise be used to grow staple food crops for local consumption, are being misappropriated to feed the already food secure? Or does the foreign exchange revenue earned from such exports, as well as the transfer of scientific knowledge and technology, genuinely improve the prospects of achieving food security? The evidence seems to suggest that many of these sites engaged in high-value agro-exports exist as production "platforms" largely disconnected from the rest of society. For example, in 2007, strawberries and flowers grown around Addis Ababa were being air freighted out of Ethiopia while other parts of the country were suffering drought and floods, and 1.36 million people were in need of emergency food aid (Bridger 2008).

On the other hand, it is important to acknowledge that around 100,000 people are directly employed in the fresh fruit and vegetable sector across Sub-Saharan Africa, with perhaps half of this total comprising small-scale farmers, the other half paid workers on large farms. At least another 100,000 people are employed in support services (Vorley 2008). This might mean that up to 1 million people in Sub-Saharan Africa are supported by high-value horticultural exports. Yet the labour practices involved leave a great deal to be desired, where women living in the

Box 6.9

Production problems in achieving export quality

Rionorte is geared to the export of melons and other fruits in the Açu Valley of North East Brazil. It enrols small farmers into its system, as well as direct farming of 11,000 ha of irrigated land. The main consideration is producing melons that satisfy the particular quality parameters of supermarkets in Europe. The firm has recently adopted integrated crop management systems at the behest of Sainsbury's [a leading UK retailer]. Considerable problems are emerging, however, concerning the intensification of production, which has proceeded at a fast pace in recent years. Farmers argue that they 'cannot keep going like this'. They argue that soils are becoming exhausted and crops are continually more prone to whitefly infestation. The cost of these disruptions in the consistency of supply is met by the farmer. These severe problems of maintaining "quality" supply, as well as the growing pressures from the retailers and the external market to increase quality parameters, has become a major source of friction between the independent producers and Rionorte. Agronomists are trying to sort the problems out, only to find that solutions seem to be temporary and beyond technological fixes.

The farmers suggest that applying herbicides and pesticides is not the solution, and new areas of production and continued intensification are not seen as sustainable. The producers see three interrelated problems that they are facing:

- the continual cost–price squeeze, with escalating cost of inputs but not farm-gate prices as these problems increase;
- the growing problems of risk of infestation from fruit and whitefly;
- the growing "quality" regulations being placed on them by Rionorte in the name of maintaining export market access.

The smaller farmers want to innovate by trying green pepper production, for instance, and also recognise that they could produce using fewer pesticide applications. But such products would not be seen as the right size, shape and skin colour "for the housewife". One farmer admitted "so I have been applying more (pesticide) to satisfy the housewife".

As Marsden concludes: "This [. . .] example shows how many small producers are getting trapped in the interstices between the skewed food quality and environmental parameters, and the development trajectories of the export enterprises which can only abide by the conceptions of the 'final consumer' and 'housewife' articulated to them by external retail capital."

Source: adapted from Marsden (2003), 58–60.

slums of Nairobi are employed in packing sheds to work 10–12 hour days (or more if an order arrives from a retailer requiring immediate shipment), six days per week. Indeed, the globalisation of food seems to go hand in hand with the need for a highly flexible labour force, sucking in cheap migrant labour to perform poorly paid field and packing tasks, whether in Lincolnshire, Almeria, California or Kenya.

Meanwhile, developing countries are urged to demonstrate reciprocity of access to their domestic markets and allow the North to dispose of its own agricultural commodities. Unfortunately, the widespread use of subsidies by the EU and the USA, particularly for export, has enabled them to flood local markets with grain at prices below the costs of production. This creates significant distortions and huge disincentives for local production of food staples, and cannot hope to improve food security, particularly where subsidised sales are likely to come with strings attached. Moreover, as we have seen, the volatility of international markets makes a hazardous business of depending upon commercial purchases for anything more than a minor proportion of food needs. This is why governments should ensure domestic food production is promoted alongside export agriculture, and that food security for all its citizens takes place alongside a commitment to create, as far as possible, a degree of food sovereignty – an issue discussed later.

Consumption

This chapter seeks to develop a broader understanding of food security than the more conventional definitions that focus upon inadequate calorific intake. While access to sufficient food remains the most pressing problem for up to 1 billion people, rapid dietary change has affected possibly twice that number, creating a process that has been labelled the "nutrition transition". The originator of this term, Barry Popkin, includes the following dietary shifts, amongst others, that have occurred across the globe:

● "The world's food supply and diets have been sweetened tremendously.
● Edible oil intake has grown very rapidly, particularly in Asia, the Middle East and Africa.
● The energy density of diets – particularly of the low-income world – seems to be growing rapidly
● The intake of animal-source foods is increasing rapidly in the low-income world." (Popkin 2005: 724–25)

As a consequence, allied with changing patterns of physical activity, the following shifts in body composition have occurred:

● "Obesity is growing rapidly on a global basis and the rate of change is faster in the low-income world.
● The burden of obesity is shifting to the poor around the globe." (*ibid.*: 725)

Popkin goes on to identify a number of factors that have contributed to this process: increased urbanisation and the creation of a more obesogenic environment; increased household income directly linked to more energy-dense diets; falling prices, especially for foods of animal origin; expansion of mass media and the role of advertising; and the globalisation of food marketing and distribution, especially the growth of global supermarket chains at the expense of traditional retailing including open markets selling local fresh produce. Although globalisation is a complex process, it is clear it is having a major effect on dietary patterns and therefore on food security.

"Globalisation", according to Hawkes (2006), "is a dynamic process of both mass global change and local differentiation." In dietary terms, she considers this as comprising the apparently contradictory elements of "dietary convergence" and "dietary adaptation". By dietary convergence, she refers to a process in those countries that are more integrated into the world economy where diets demonstrate "increased consumption of meat and meat products, dairy products, edible oil, salt and sugar, and a lower intake of dietary fibre". Dietary adaptation, in contrast, is "increased consumption of brand-name processed and store-bought food, an increased number of meals eaten outside the home and consumer behaviours driven by the appeal of new foods", reflecting increased exposure to advertising and the emergence of new retail outlets.

While globalisation has often been represented in dietary terms as a purely homogenising process – represented elsewhere as "McDonaldisation" or "Coca-Colonisation" – Hawkes is at pains to stress that the two dynamic and competitive forces of convergence and adaptation create more complex outcomes. Nevertheless, these dietary outcomes are associated with

> rising rates of overweight, obesity and diet-related chronic diseases, like heart disease, diabetes and some cancers. More people now die of heart disease in developing countries than in developed, and the problem is becoming more serious among the poor. Low quality diets are also associated with undernutrition in the form of micro-nutrient deficiency,

which ... lowers immunity to infectious diseases. Poor quality diet is thus associated with a dual burden of malnutrition and disease.

(*ibid.*: 2)

Hawkes provides some revealing illustrations of how globalisation and trade liberalisation, involving investments by transnational food companies into developing countries, have transformed dietary patterns. Box 6.10 provides a summary of her Mexican case study.

It is evident from the discussion of both production and consumption aspects that policies designed to integrate the global food market – in agriculture, trade and foreign direct investment – developed entirely in the economic sphere have enormous influence on dietary habits. They are not just global *economic* policies, but effectively global *health* policies. In this, transnational food companies play an influential role: directly, by affecting dietary habits through the manufacture, promotion and retailing of processed foods; and indirectly, by altering the parameters of domestic food markets through the introduction of new cultural markers. While globalisation is a complex, multidimensional process from which it is hazardous to draw singular outcomes for food security, it is nevertheless encouraging the uneven development of dietary habits between rich and poor. Here, the more affluent are able to use their higher disposable incomes to access healthy and "diet" foods within a more international cuisine, while the poor make the dietary transition from a possibly restricted range of culturally traditional staples to processed, energy-dense foods high in trans fats, salt and sugar. While welcoming the chance to widen their diet, it is the latter group that consequently carries the burden of obesity and associated ill health.

7. Food sovereignty

Food sovereignty is the right of peoples to define their own food and agriculture; to protect and regulate domestic agricultural production and trade in order to achieve sustainable development objectives; to determine the extent to which they want to be self reliant; to restrict the dumping of products in their markets, and to provide local fisheries-based communities the priority in managing the use of and the rights to aquatic resources. Food sovereignty does not negate trade, but rather, it promotes the formulation of trade policies and practices that serve the rights of peoples to safe, healthy and ecologically sustainable production.

(Forum for a New World Governance 2010)

Box 6.10

Foreign direct investment in the manufacture and retail of processed foods

The globalisation of the Mexican food economy is profoundly linked with its neighbour, the United States. Under the North American Free Trade Agreement (NAFTA), which came into effect in 1994, there was rapid acceleration of FDI from the USA into Mexican food processing. In 1998, sales from US food industry affiliates in Mexico exceeded US$12 billion, easily surpassing the value of US processed food exports (US$2.8 billion).

Between 1995 and 2003, sales of processed foods expanded by 5–10 per cent per year in Mexico. Consumption of Coca-Cola drinks rose from 275 8 oz servings per person per year in 1992 to 487 servings in 2002 (greater than the 436 servings per person in the USA). Eating *comidas chatarras* (junk food) is very common among children in parts of the country. Higher consumption of these energy-dense foods is thought to be associated with increased consumption of dietary fats and sugars in Mexico, which has in turn been linked with obesity and diet-related chronic diseases. Concern about obesity and diabetes is in fact leading to a counter-trend in the processed foods market: increased sales of "diet" foods.

A second effect of NAFTA on the processed foods market was to stimulate the growth of multinational retailers. This has resulted in an explosive growth of chain supermarkets (Walmart is now the leading retailer and employs more people than any other company in Mexico); discounters; and convenience store chains (in which market leader OXXO, owned by a Coca-Cola subsidiary, tripled its number of stores between 1999 and 2004). By 2004, in just a decade since NAFTA, supermarkets, discounters and convenience stores accounted for 55 per cent of all food retail in Mexico, and now dominate the sector in large and medium-sized cities. Yet even the small, family-owned *tiendas* that predominate in small towns and rural areas play an important role in the expansion of the processed foods market in Mexico.

Obesity is now quite high in some poor rural communities, with the greatest relative changes occurring in the poorer southern region. Trends in obesity and diabetes are reaching "epidemic" proportions, with the rate of overweight/obesity estimated at 62.5 per cent in 2004, while 8 per cent of Mexicans have diabetes, costing the country US$15 billion a year.

Source: adapted from Hawkes (2006).

Food sovereignty is most closely associated with civil society organisations (CSOs) and social movements engaged in the struggle against globalisation, but in recent years is a term that, if not quite mainstream in Washington, has certainly entered the vocabulary of agri-food policy analysts and advisors. It offers a counter-hegemonic perspective on food, rooted in a rights-based framework that effectively insists upon food being treated as a basic human right. It has been widely proclaimed and reaffirmed at meetings and forums held in parallel with events such as the World Food Summit of 1996 and its follow-up in 2001, and a host of other gatherings around the world. Although it has become the widely adopted slogan of a broad-based and non-hierarchical movement, food sovereignty is most closely associated with the CSO La Vía Campesina ("peasant way" in Spanish), the International Movement of Small Farmers.

Food sovereignty is largely formulated as an alternative policy proposition to liberalised industrial agriculture and is based upon a number of core assumptions. First, it attaches almost primordial significance to the family farm, which itself is located within a community-based rural development model. Clearly, this has to be underpinned by access to sufficient land, so agrarian reform features as one of La Vía Campesina's core principles by which the landless and marginalised are given ownership and control of the land they would work. While enormous disparities in landholding do exist in many countries, and agrarian reform might help to improve the efficient use of land, experiences of reform in other countries have demonstrated that reallocation of land is no guarantee of food security (in the absence of tools, seeds, water, etc.). Increasingly, there is a need to rethink land tenure institutions beyond individual property rights, with forms of common pool resources management offering a more collective solution while ensuring greater social inclusion and equity. Besides, would agrarian reform involve the parcelisation and distribution of land currently performing vital ecosystem services (e.g. tropical forests) in order that all might establish family farms?

Yet a second assumption suggests a negative response to this question, for there is some emphasis attached to sustainable methods of production, utilising indigenous biodiversity (seeds and livestock breeds) and reducing dependence upon agri-chemicals. Here, much greater attention is placed upon utilising farmers' existing agricultural knowledge and locally adapted technologies. The term "agro-ecology" often features as shorthand to denote a wide range of practices that have built upon tried-and-trusted indigenous methods and that operate in tandem with local resource constraints and possibilities. Interestingly, while critics of such an approach might argue that only the most modern technologies can offer a future of greater food output, recent reports have tended to be much more cautious in proclaiming the advantages of the latest seeds and higher levels of inputs. The recently published Synthesis Report (IAASTD 2009) makes interesting reading in this context (Box 6.11).

A third and, for our purposes here, final assumption drawn from those that underpin the notion of food sovereignty concerns its very

---

Box 6.11

Rethinking agricultural knowledge, science and technology

International Assessment of Agricultural Knowledge, Science and Technology for Development (IAASTD) responds to the widespread realisation that, despite significant scientific and technological achievements in our ability to increase agricultural productivity, we have been less attentive to some of the unintended social and environmental consequences of our achievements. We are now in a good position to reflect on these consequences and to outline various policy options to meet the challenges ahead, perhaps best characterised as the need for food and livelihood security under increasingly constrained environmental conditions from within and outside the realm of agriculture and globalised economic systems. For many years, agricultural science focused on delivering component technologies to increase farm-level productivity, where the market and institutional arrangements put in place by the state were the primary drivers of the adoption of new technologies. The general model has been continuously to innovate, reduce farm-gate prices and externalise costs. This model drove the phenomenal achievements of agricultural knowledge, science and technology (AKST) in industrial countries after the First World War and with the spread of the Green Revolution beginning in the 1960s. But, given the new challenges we confront today, there is increasing recognition within formal science and technology organisations that the current AKST model requires revision. "Business as usual" is no longer an option. This leads to rethinking the role of AKST in achieving development and sustainability goals; one that seeks more intensive engagement across diverse worldviews, and possibly contradictory approaches in ways that can inform and suggest strategies for actions enabling the multiple functions of agriculture.

Source: adapted from IAASTD (2009), 3.

proclamation of *sovereignty* in a globalised world. In this respect, it appears as both defender of the nation state, as constituting the sole legitimate authority to determine policies that affect its people; and critic of the globalisation project in general and its key agencies in particular. The nation state has certainly lost ground in exercising power to regulate many aspects of food and agriculture, a process that began with the implementation of structural adjustment programmes during the 1980s. From this moment, the IMF and the World Bank began to assert more direction over macro-economic policy in many countries of the South, imposing conditionalities on loans in order to encourage countries to open up their markets to foreign goods and services. It is at the creation of the World Trade Organization (WTO), however, that proponents of food sovereignty have directed much of their ire.

The WTO came into existence in 1995 following protracted international negotiations known as the Uruguay Round. Although agriculture and food were one of the most divisive issues within these negotiations, particularly between the USA and the EU, sufficient grounds were established to create the Agreement on Agriculture (AoA). This was designed to "improve market access and reduce trade-distorting subsidies in agriculture" with all signatory countries committed over a six- (North) or ten- (South) year period to eliminate tariffs, subsidies and other domestic supports in favour of free and open markets. Within Europe, the Common Agricultural Policy exists to support European agriculture and, despite a reform process that remains a highly contested and controversial issue between EU member states, still accounts for over 40 per cent of the total EU budget. In other words, critics believe that there remains fundamental inequality in the operation of the AoA, as the EU and USA continue to deploy domestic supports for their farmers but insist on their removal in the rest of the world.

Besides campaigning for the removal of agriculture from the WTO, there are a host of other agreements that have a major bearing on food sovereignty. These include Agreements on Trade Related Intellectual Property Rights (TRIPs); Sanitary and Phytosanitary (SPS) measures; Technical Barriers to Trade (TBT); Subsidies and Countervailing Measures (SCM); and the General Agreement on Trade in Services (GATS). According to critics of the WTO, all these agreements strengthen the power and control of transnational corporations, such as seed companies seeking to extend patent rights over plant genetic resources nurtured and developed by generations of farmers in the South. In short:

> Global trade (rules) must not be afforded primacy over local and
> national developmental, social, environmental and cultural goals.
> Priority should be given to affordable, safe, healthy and good quality
> food, and to culturally appropriate subsistence production for domestic,
> sub-regional and regional markets.
>
> (Forum for a New World Governance 2010)

Food sovereignty, then, is not simply another definition of food security,
but provides a radical challenge to many of the existing assumptions
about the way food and agricultural policies have developed, and might
continue to be developed. Its perspective is not that of the academy or of
those in FAO headquarters in Rome, but of the rural poor, hungry and
food insecure. As Windfuhr and Jonsén (2005) note, there is no single
fully fledged food sovereignty model with a set of policies available for
governments to implement. Yet, although there will be many vested
interests deeply and violently opposed to much of what the notion
represents, it is being developed by CSOs and social movements all over
the world to improve the governance of food and agriculture and to
address the core problems of hunger and food insecurity.

8. Summary: rethinking food security?

This chapter contains a broad-brush discussion of food security.
Beginning with the conventional preoccupation with hunger and
malnutrition in the South, it explores the inseparable association with
food poverty, a concern of social policy analysts in the developed world.
Together, these twin concerns highlight calls for social justice, human
rights and community empowerment. The chapter explores some of the
significant challenges for global food security: population growth and
changing patterns of consumption; peak oil and energy security; climate
change; and globalisation. It should be noted that these are not the only
global challenges to food security: to this list we might have added the
prospective scarcity of freshwater, the continuing loss of topsoil and other
forms of land degradation, and the erosion of biodiversity and ecosystem
support services, issues discussed in chapter three. In other words, the
circumstances facing the achievement of global food security in the
decades ahead promise to be more difficult than even those of the past.

Yet, if there has been such a failure to meet the target of halving the
number of hungry by 2015 set by the World Food Summit of 1996 and
reiterated by the Millennium Development Goals, perhaps this calls into
question the effectiveness of the existing organisational architecture of

the world food system. Despite the high-level conferences on world food security, such as the one held in Rome in June 2008, and the formation of a High-Level Task Force on the Global Food Crisis chaired by the UN Secretary-General, charged with catalysing urgent action (FAO 2008), it may be that food security needs rather less global leadership and more local-level action. For, arguably, it is at the local level where the notion of food security is best grounded: how to achieve access to adequate food that is culturally and nutritionally appropriate throughout the year, and from year to year, that provides for health and wellbeing. This might require fewer "high-level" events, instead using the resources to enhance access to micro-credit by small farmers and rural producers.

Such an approach would embrace more publicly funded, rather than privately led, investment in agricultural research, where less emphasis would be placed on finding a magic bullet associated with gene technology, and more on building adaptive capacity, resilience-enhancing systems of production, and locally appropriate technology portfolios. Finding ways to improve adaptation will be the key to building food and other dimensions of human security within a warmer, more crowded and more complex world.

This approach would necessarily rework understandings of food security, including those derived from specific local circumstances, and embark from a commitment to social justice, environmental sustainability and sound nutrition. It might be that food security would be facilitated, *contra* von Braun and Diaz-Bonilla (2008), by less, rather than more, globalisation. It might go further and argue for food sovereignty – effectively the right of local farmers to grow food for local consumers, rather than exclusively agri-commodities for export. Without retreating into autarchy, it might enshrine the basic principle that each country should endeavour to produce enough food to feed its own people. While this may seem a radical set of measures, it is apparent that trade liberalisation in food and agriculture has not delivered global food security to date, and that the diverse challenges ahead should be a cause to reflect upon a change of direction.

Further reading

Lawrence, G., Lyons, K., Wallington, T. (eds) (2010) *Food Security, Nutrition and Sustainability*. London: Earthscan.
Interesting and wide-ranging, recently published collection of essays that breathes new life into the concept of food security.

Brown, L. (2005) *Outgrowing the Earth: The Food Security Challenge in an Age of Falling Water Tables and Rising Temperatures.* London: Earthscan.

An indefatigable environmentalist with the capability to harness huge amounts of data in support of an argument that I do not always entirely share – but an extraordinarily important perspective nonetheless. Vital work.

Holt-Giménez, E., Patel, R. (2009) *Food Rebellions! Crisis and the Hunger for Justice.* Cape Town: Pambazuka Press.

This is *not* the FAO view of food security. Bombastic but invigorating take on the causes of food insecurity and how to address it.

Windfuhr, M., Jonsén, J. (2005) *Food Sovereignty: Towards Democracy in Localized Food Systems.* Bourton-on-Dunsmore, UK: ITDG Publishing. www.ukabc.org/foodsovpaper.htm

And on the web

Forum for a New World Governance (2010) 'People's Food Sovereignty Statement', www.world-governance.org/spip.php?article70

People's Coalition on Food Sovereignty: www.foodsov.org

La Vía Campesina: http://viacampesina.org/en

The International Planning Committee for Food Sovereignty: www.foodsovereignty.org

7 Towards a sustainable agri-food system

Contents

1.	**Introduction**
2.	**Reconsidering sustainability**
3.	**Sustainable agriculture**
4.	**Sustainable consumption**
	Fairtrade
	Reducing the consumption of meat
5.	**Reconnecting production and consumption for sustainability**
	Farmers' markets
6.	**Planning and creating sustainable food systems**
	Urban agriculture
	Public procurement of food
	Growing sustainability from the roots?
7.	**Summary**
	Further reading

Boxes

7.1	Sustainability as functional integrity and relative equilibrium
7.2	Sustainability as dynamic, complex and contested
7.3	Agricultural sustainability
7.4	Integrated pest management in rice
7.5	Fairtrade and market sovereignty
7.6	A typology of short food supply chains
7.7	School food in East Ayrshire
7.8	Community food security in action
7.9	Growing local food

1. Introduction

It has become clear during the course of this book that the contemporary global agri-food system – despite its achievements in producing more calories for less money than ever before – is facing a number of profound challenges. Each of the major components of the system is associated with a particular set of problems, although we have seen how dependence on fossil fuels and contributions to global warming are common features throughout. For primary food production, where industrialised methods have come to dominate across large parts of the world, key issues include the excessive utilisation of soil and water resources, the loss of ecological services, the depletion of biodiversity (including wild fish stocks), and anxieties around the safety and ethical dimensions of intensive livestock production. For food processing and manufacturing, which encompass a wide range of activities, the drive to deliver convenience and low prices has resulted in many food products deleterious to nutritional health, and packaging waste streams that threaten to overwhelm the capacity of local infrastructures. Third, corporate retailing, which has become the gravitational centre of the agri-food system, is exerting extraordinary influence, both upstream through its supply chain and the contracting of producers, and downstream by shaping the purchasing behaviour of consumers, including the buying of more than is needed. Waste and overconsumption are key consequences of a system designed to maximise throughput and opportunities for profit. Finally, and most unforgivably of all, while the majority of people in the North and many in middle-income countries enjoy the luxury of abundance, choice and low food prices, over 1 billion people in the world today are hungry and, for at least part of the year, malnourished.

Viewing the agri-food system as a whole, it is apparent that it is marked by a profound disconnection between consumers of food and sites of primary production; by high externalities in both environmental and human health terms; and by high levels of inequality of access to adequate nutrition. It is rational to argue that the global agri-food system is unjust, wasteful and utterly unsustainable in its current form, and requires thorough restructuring not only to address these failures, but also to make it capable of adapting to the stresses of climate change, water scarcity and rising energy costs. This chapter explores the possibilities for creating a more socially inclusive and environmentally sound food system. The first section briefly explores the meaning of sustainability as it might be applied to the agri-food system, and subsequent sections deal with sustainable primary production, sustainable consumption, the need to reconnect production and consumption, and, finally, the planning and creation of a sustainable agri-food system.

2. Reconsidering sustainability

At the beginning of this book, the notion of sustainability is briefly introduced as a composite term capable of conveying important principles – including productivity, stability, equity and diversity – across biological, economic and social realms. It is suggested that one of the key objectives of sustainability is to optimise goal achievement across these three realms through an adaptive process of trade-offs that is place- and time-specific.

It should be clear at this stage of the book that there is little about the contemporary global agri-food system that can be regarded as sustainable: the pursuit of productivism has occurred at the expense of natural resources, rural economies and social wellbeing, rather than being traded off to ensure optimal outcomes across all realms.

Consequently, in this chapter we need to ask whether there is any prospect at all to shift the production, processing, distribution, retailing and consumption of food onto a more sustainable basis. Before answering this question, it is important to establish that sustainability is not a formulaic notion that can be straightforwardly defined. Nor should it be regarded as constituting a relatively fragile and elusive equilibrium that we struggle to secure and maintain. This is a perspective that is widely held, and Box 7.1 offers an eloquent example in which sustainability is viewed as sets of interconnected subsystems held together within a functionally integrated balance of nature.

Thompson's observations make clear the multiplicity of interlocking subsystems that contribute to the performance of the whole system. Even if any one of those subsystems is operating suboptimally, he cautions about the dangers of tampering with it for fear of destabilising the whole. This idea of sustainability as constituting a steady state, where all elements are held together within a notional "balance of nature", has been subject to critical scrutiny. Arising from work in ecology, particularly that associated with the distinguished systems ecologist C.S. Holling, notions of resilience and stability have become more widely employed in a wide range of disciplines and have served to

---

Box 7.1

A view of sustainability as functional integrity and relative equilibrium

Consider how food production affects our society's ability to reproduce itself. This is a problem of almost overwhelming complexity, because society must be understood as a system comprising many subsystems that are threatened in different ways by different approaches to producing food. The human population's biological need for food sets one system parameter; but, in meeting this parameter, it is possible to deplete soil, water and genetic resources used in food production. Since each of these is a regenerative subsystem, threats to these subsystems represent threats to total system sustainability. Similarly, farms and rural communities represent subsystems. If farming is unprofitable, or if the local institutions that support farming are not regenerated, the sustainability of the larger system is threatened. Our desire to maintain the functional integrity of all these subsystems might make us cautious about tampering with any subsystem that seems to be functioning, for fear that what we do might upset the complex interconnection of the whole.

Consequently, far from understanding sustainability as one dimension of optimisation, we would understand it as a *relative equilibrium* among social and natural subsystems, an equilibrium that we challenge at our peril.

Source: adapted from Thompson (2010), 22 [my emphasis].

---

inform the idea of "complex adaptive management of social–ecological systems". Whereas a static equilibrium view, as represented by Thompson's extract, would seek to control change, a resilience perspective would promote policies and interventions that enable systems to cope with, adapt to, and even shape change. As Folke (2006: 254) observes, "managing for resilience enhances the likelihood of sustaining desirable pathways for development in changing environments where the future is unpredictable and surprise is likely".

Given the range and scale of the environment and development challenges faced by the global agri-food system, and the failure of so many policy interventions to date (such as resolving hunger), it appears that this approach to sustainability might help in shaping a different kind of future. This would have to start by rethinking prevailing assumptions. These include the privileged role attached to the prevailing scientific method that seeks "magic-bullet", technological solutions to complex, shifting and dynamic problems; the tendency to design standardised "blueprint" models that are somehow applicable to many different contexts; and the assumption that "progress" everywhere constitutes an unproblematic linear trajectory toward higher productivity. This approach, in contrast to the equilibrium view of Thompson and others, places greater store on complexity, dynamism and resilience. It alerts us to the multiplicity of interactions between different systems (social, ecological, technological, economic) and across multiple scales (global, national, local), but does not regard them fearfully as needing to be stabilised, regulated and controlled. Rather, it understands that complex social–ecological systems will be continuously reshaped by global forces, but also by local circumstances, local capabilities and local knowledge. And in each place, different social groups, each with their own set of values and resources, will frame notions of "progress" or wellbeing in quite different ways. Aspects of this perspective are developed in Box 7.2.

Box 7.2

Sustainability as dynamic, complex and contested

- Conventional approaches to development and sustainability have often been rooted in standard equilibrium thinking. This tends to centre analyses on attempts to control variability, rather than adapt and respond to it. Conventional methods often assume that models developed for one setting – usually the more controlled – will work in others; thus the export of models from the developed to the developing world, or from the laboratory to the field.
- While governments and institutions are increasingly preoccupied with risk and with insecurities that real and perceived threats seem to pose, dominant approaches involve a narrow focus on a particular and incomplete notion of risk. This assumes that complex challenges can be calculated, controlled and managed. But some situations involve *uncertainty*, where the possible outcomes are known, but there is no basis for assigning probabilities. Other situations involve *ambiguity*, where there is disagreement over the nature of the outcomes, or different groups prioritise concerns that are incommensurable. Finally, some situations involve *ignorance*, where we don't know what we don't know, and the possibility of surprise is ever present. Whereas expert-led approaches may be well attuned to handling risk, they are inadequate in increasingly common situations where other kinds of incomplete knowledge can be recognised to prevail.

- Underlying such approaches are often wider assumptions about what constitutes the goals of "development" or "sustainability", often assuming a singular path to "progress" and a singular "objective" view of what the problem might be. Yet different people and groups often understand system functions and dynamics in very different ways. They bring diverse knowledge and experience to bear. People also value particular goals and outcomes in very different ways. Consequently, we can increasingly recognise situations in which there are a multiplicity of possible goals, which are often contested.
- While debates about sustainability have become mainstream over the past two decades, they have also given rise to confusion and fuzziness, in which rhetorical use masks lack of real change and commitment. Ideas of sustainability have become co-opted into inappropriately managerial and bureaucratic attempts to "solve" problems that are far more complex. Rather than using sustainability in a general sense to imply the maintenance of (unspecified) features of systems over time, it should be used to refer to explicit qualities of human wellbeing, social equity and environmental integrity, and the particular system qualities that can sustain these. These goals are context-specific and inevitably contested, meaning that it is essential to recognise the roles of public deliberation and negotiation – in defining both what is to be sustained and how to get there – in what must be seen as a highly political (rather than technical) process.

Source: adapted from Leach *et al.* (2010), 3–5.

It is clear that, in following the approach outlined by Leach *et al.*, we are moving away from simplified definitions of sustainability and into a field that is complex, contested and contingent. Such an approach embarks from an understanding of systems that are prone to both longer-term stresses and more transient but disruptive shocks. These might originate as agro-ecological problems (pests, diseases), bio-physical hazards (drought, floods), or trends in resource depletion (freshwater scarcity) or global environmental change (climate warming, drying). They might, on the other hand, occur in an entirely different realm, such as volatility in international market prices detrimentally affecting farmers or consumers. In whatever form change occurs, it is vital that the prevailing system is capable of responding to such challenges. In this regard, Leach *et al.* (2010) outline four dynamic system properties – the ability of a system to sustain structure by:

- controlling sources of short-term episodic shocks (*stability*);
- responding effectively to short-term episodic shocks (*resilience*);
- controlling sources of long-term stress (*durability*); and
- responding effectively to long-term and enduring stress (*robustness*).

The key to distinguishing between these four system properties is whether a particular system has the capacity to *control* a shock or stress by intervening to eliminate its cause or consequence, or must *adapt* to it through its own internal capacity to evolve and change. The notion of complex adaptive systems is becoming especially important in thinking about sustainability, for it is likely that the challenges ahead will require all social–ecological systems to demonstrate flexibility. In applying this thinking to the study of agri-food systems, then, it becomes apparent that sustainability takes on a very important role, requiring us to evaluate the resilience or durability of different approaches to agricultural development and their capacity to withstand stresses and shocks. It also reminds us that social diversity and differential access to benefits will result in many different prioritisations of available technological choices.

Consequently, in the context of current calls "to double food production in order to feed 9 billion people by 2050" (e.g. in a statement made by Jacques Diouf, Director-General of the UN Food and Agriculture Organization, FAO, at the high-level conference on world food security in June 2008), and championed by those who represent the most powerful global agri-food interests, it is necessary to question whether productivism is really the most appropriate response. Faced with an increasingly complex landscape involving pressing environmental, social justice and economic development challenges, new approaches informed by sustainability principles must be explored. For the drive to double food production will not necessarily result in significantly fewer food insecure and hungry people. This is because, as chapter six makes clear, the matter of hunger has less to do with availability of food and more to do with access and entitlement. Biotechnology has not yet promised to resolve these issues, nor is it likely that it will be in a position to do so. For this and many other reasons, the case for a more sustainable approach to primary food production, particularly agriculture, is increasingly coming to the fore.

3. Sustainable agriculture

Despite its achievements in raising food output, productivist agriculture has resulted in a significant depletion of the resource base on which it depends. Yet critical voices who propose alternative forms of production that make better use of ecological services, regenerative processes and local farming knowledge are frequently drowned out by those anxious to defend the *status quo* or to protect dominant commercial interests, or simply fearful that "lower-tech" methods would not be able to match current output volumes. It should be noted that the overwhelming effort in agricultural research and extension has been focused upon mainstream productivism, an effort that stretches from the international centres that form part of the Consultative Group for International Agricultural Research (CGIAR), through publicly funded national agencies, to a range of private-sector institutions (especially agri-chemical, seed and biotechnology interests) with university research centres often straddling public and private spheres. Much of this work, though by no means all, has focused upon the creation in research labs of such "magic-bullet" technologies as genetically modified (GM) seeds where, after thirty years, just two traits in four crops (one of which is designed to allow plants to tolerate a proprietary herbicide) in six countries account for 95 per cent of the area sown to GM (EC SCAR 2009). In contrast to this powerful alliance of "high-tech" interests, there has been comparatively little funding to support scientific research on more sustainable alternatives. Yet such work has been undertaken in rural communities worldwide, invariably involving farmers as subjects of their own development, not as passive recipients of technologies developed elsewhere.

Such interventions demonstrate an entirely different worldview from that of productivism and engage with a bigger challenge than devising a technology to increase yield. That challenge for primary food production worldwide is not just how to increase productivity, but how to do so in ways that ensure more sustainable use of ecological services and natural resources while delivering balanced nutrition to all peoples of the world. This challenge requires a fairly profound rethinking of established norms across all aspects of conventional agriculture. Fortunately, the Synthesis Report of the International Assessment of Agricultural Knowledge, Science and Technology for Development (IAASTD 2009) reveals that such rethinking is under way. Indeed, the IAASTD process represents something of a sea change in policy terms. Constituted as a UN-sponsored scientific consultation exercise (much like the Millennium Ecosystem Assessment) that brought together over 400 scientists, it made a series of recommendations that have been approved by sixty-one governments (though not those of the USA, Canada and Australia), and welcomed by grassroots food and farming groups, although rejected by the biotechnology industry and effectively ignored by the FAO and the World Bank (Holt-Giménez and Patel 2009). It demonstrates, however, that the case for sustainable primary food production systems is being made, as the brief extract in Box 7.3 illustrates.

Nevertheless, there is some work yet to do in order to change the prevailing paradigm dominated by a preoccupation with agricultural productivity. Fernandes *et al.* (2005) helpfully capture this worldview by setting out four equations that have come to dominate agricultural research and extension and which, they argue, are in need of revision. They are:

● control of pests and diseases = application of pesticides or other agrochemicals;
● overcoming soil fertility constraints = application of chemical fertilisers;
● solving water problems = construction of irrigation systems;
● raising productivity beyond these three methods = genetic modification (*ibid.*: 329).

As these authors explain, it is not that these equations are wrong – they have, after all, produced some impressive results in productivity terms – rather, they have become utterly dominant in establishing and maintaining structures that determine what is regard as acceptable, or orthodox, in the practice of agricultural development. Fortunately, there are a host of established practices that offer solutions to each of the problems posed by the equations, and often do so at lower cost as well

---

Box 7.3

Agricultural sustainability

Agricultural knowledge, science and technology (AKST) systems are needed that enhance sustainability while maintaining productivity in ways that protect the natural resource base and ecological provisioning of agricultural systems. Options include improving nutrient, energy, water and land use efficiency; improving the understanding of soil–plant–water dynamics; increasing farm diversification; supporting agro-ecological systems and enhancing biodiversity conservation and use at both field and landscape scales; promoting the sustainable management of livestock, forest and fisheries; improving understanding of the agro-ecological functioning of mosaics of crop production areas and natural habitats; countering the effects of agriculture on climate change and mitigating the negative impacts of climate change on agriculture.

Source: adapted from IAASTD (2009), 5–6.

as with greater environmental and social benefit. The challenge, then, is not finding new solutions to established problems, but changing the structures and mindset that maintain these narrowly conceived propositions. Let us briefly examine the first two equations.

First, in relation to the control of pests and diseases, there is probably no better illustration of mechanistic thinking in agricultural production than when farmers are advised to spray on a routine calendar basis: "spray at weekly intervals" is a powerful extension message (Pretty and Hine 2005). Although applications may be necessary on occasion, the fact that it has become the action of first resort rather than when levels of pest infestation exceed an economic threshold illustrate the way in which farmers' own judgement can be overwhelmed by privileged, but not necessarily appropriate, technical expertise from the outside. In the North, where a much higher level of environmental and health and safety regulation prevails, strict controls apply to the availability of certain agri-chemical products and their use. In the South, on the other hand, particularly in areas where there are strong commercial incentives to intensification, highly toxic agri-chemicals (insecticides, fungicides, herbicides, rodenticides) are usually widely available, depending more on the distribution networks of manufacturers and suppliers than on the regulations of the state. Here, acute pesticide poisoning is a serious public health issue and has resulted in a significant legacy of human morbidity and mortality and ecological disruption (Kishi 2005).

Pesticides can be dangerous to human health and damage natural resources but, more importantly to the farmer, pesticides are often inefficient at controlling pests. They can cause pest resurgences by killing off the natural enemies of the target pests. They can produce new pests by killing off the natural enemies of species that hitherto were not pests. Pests and weeds can also become resistant to pesticides, so encouraging further applications. And pesticides provide no lasting control and so, at best, have to be applied repeatedly (Pretty 1995; Pretty and Hine 2005).

Fortunately, the control of many different kinds of pests and diseases has been proven time and again through recourse to integrated pest management (IPM) practices. These vary according to the nature of the problem, but can include the encouragement of beneficial and predator species such as parasitic wasps, spiders and beetles; the use of "natural" pesticides that are locally available as plants with insecticidal properties or disease-controlling properties (e.g. chilli pepper, tumeric, neem); the release of pheromones that disrupt the mating behaviour of targeted insects; or the deployment of a range of cultural practices such as rice–duck or rice–fish farming. Some of the most advanced and sophisticated IPM strategies are deployed in Asian rice-growing systems, in light of the growing realisation that heavy applications of pesticide to eliminate the rice brown planthopper gave rise to worse outbreaks of infestation. This is described in Box 7.4.

In relation to the second equation, that soil fertility constraints may be overcome only with recourse to chemical fertilisers, there is also growing evidence to demonstrate the applicability and success of alternative approaches. Invariably, it helps to begin by rethinking the "problem" rather than simply applying a singular "solution". In the case of soils, for example, chapter three notes how the rise of productivism led to increasing reliance on the three macro-nutrients – N, P, K – and a consequent neglect of the role of organic matter and the natural cycling of nutrients. The application of inorganic fertilisers enhanced the scaling-up of production, given the ease of supplying plant nutrition in a

---

Box 7.4

Integrated pest management in rice

"A particular characteristic of Asian rice ecosystems is the presence of a potentially damaging secondary pest, the rice brown planthopper (BPH). This small but mighty insect has in the past occurred in large-scale outbreaks and caused disastrous losses. These outbreaks were pesticide induced and triggered by pesticide subsidies and policy mismanagement. BPH is still a localised problem, especially where pesticide overuse and abuse is common, and therefore can be considered as an ecological focal point around which both ecological understanding and management are required for profitable and stable rice cultivation.

[Since the introduction of IPM] the first, and perhaps strongest indicator, is the greatly reduced incidence of BPH. Wide area outbreaks accompanied with massive losses have no longer been experienced during the past 15 years since IPM programmes have become widely implemented in both policy and field training. In most cases, changes in policy involved removal of pesticide subsidies, restrictions on outbreak-causing pesticides, and investment in biological research and educational programmes for decision makers, extension and farmers. These policy changes most often came about as a result of successful small-scale field trials."

Source: Gallagher *et al.* (2005), 207–8, 214.

bag, and led to impressive improvements in yield. But many soils, especially in the tropics, are not well suited to this simplified, industrial-style input–output arrangement, and even previously good soils have experienced a significant loss of fertility and structure with deterioration in the complexity and function of the soil food web.

Rethinking the soil fertility equation consequently requires breaking with the application of *more* chemical fertilisers as the singular solution to overcoming fertility constraints. It means embarking upon a reappraisal of soils using a *biological* rather than a chemical approach, finding ways of increasing organic matter, microbial activity and nutrient cycling. Thinking biologically may not offer the simplicity of identifying chemical deficiencies and supplying quick-fix remedies, for it requires a more holistic appraisal of the multiple dimensions of good soil health as the basis for long-term crop productivity. Invariably, this requires rethinking cropping systems, including challenging the prevailing notion that farming must involve continuously cultivated monocultures: "clean-ploughed fields, sown uniformly in a single crop, planted neatly in rows with all extraneous plants removed" (Fernandes *et al.* 2005: 328). For the growing body of evidence is that yields, yield stability, nutritional quality per unit of land, and soil quality are superior in polyculture systems involving a combination of plants, including food and non-food cover crops, together with mulches, green manures and other nutrient-cycling methods.

These two examples – managing pests and diseases, and soil fertility enhancement, both without reliance on chemicals – demonstrate not only that more sustainable technologies and practices exist, but that they generally outperform industrialised productivist methods especially when evaluated in terms of resource utilisation, resilience to external shocks and stress, human and environmental health, and even with regard to total food output. Agricultural sustainability consequently seeks to combine three core principles, as follows:

● *Ecologically sound*: integrating natural processes such as nutrient cycling, nitrogen fixation, soil regeneration and pest management and by minimising the use of non-renewable inputs (pesticides and fertilisers).
● *Economically viable*: that farms produce sufficient food output capable of supporting the livelihoods of those engaged in production, but also recognise their multifunctional role as environmental stewards and as economic actors within local and regional economies.

● *Socially just*: those who produce food have rights to land, to appropriate technical support and to market opportunities; and this enhances social capital, self-reliance and growing cooperation in pursuit of food security for all.

These elements of a sustainable agriculture are common to a range of initiatives that have grown enormously in number over the past quarter-century all around the world, in North and South, from some of the richest to some of the poorest countries. Although the term "agro-ecology" is widely used, other labels are also applied to these sustainable agricultural initiatives, including low external input and sustainable agriculture (LEISA); organic or biological farming; and a variety of more "alternative" approaches including biodynamic agriculture and permaculture. For Thompson and Scoones:

> All of them, representing thousands of farms and farming environments, have contributed to an understanding of what sustainable agri-food systems are, and each of them shares a vision of "farming with nature", an agroecology that promotes biodiversity, recycles plant nutrients, protects soil from erosion, conserves water, uses minimum tillage, and integrates crop and livestock enterprises on the farm.
>
> (Thompson and Scoones 2009: 7)

The case for a more sustainable agriculture is therefore becoming more widely recognised as the challenge is not simply to produce more food, but to optimise it across a far more complex landscape of production, rural development, environmental and social justice outcomes (Pretty *et al.* 2010). Putting sustainable agriculture – and other forms of primary food production – into practice is consequently likely to lead to an enormous diversity of context-specific pathways that will seek to combine the particular social, technological and ecological elements prevailing within each area. Moving toward more heterogeneous production strategies will help to ensure greater resilience of agri-food systems, in which desired outputs are achieved without depleting natural capital and while sustaining the functioning of ecosystem services, and where the livelihoods of producers and their legitimate claims for food sovereignty are also secured. Finally, sustainable agriculture has to form part of a more sustainable agri-food system in which the physiological and cultural needs, nutritional security and dietary wellbeing of food consumers are also addressed. This requires us to think through the issues and responsibilities surrounding sustainable consumption.

4. Sustainable consumption

If the term sustainable development is regarded as an oxymoron (Redclift 1994), then *sustainable consumption* is arguably even more so, bringing together a word that means "to use up or destroy" with its complete opposite, sustaining (Peattie and Collins 2009). Sustainable consumption has existed as an idea since the UN Conference on Environment and Development held in Rio de Janeiro in June 1992. Yet it remains only weakly developed in the international policy arena – although great store is set by the aspiration to achieve it – and arguably this is due to the almost universal support for the imperative of economic growth. Without questioning, this central axiom of modern society, which largely rests upon the desirability of individual citizens earning more income with which to buy more things and enabling national economies to grow, sustainable consumption remains little more than a stepchild of ecological modernisation. Though itself a focus of considerable intellectual investment and potential complexity, a very simple definition of ecological modernisation might be to suggest that it seeks to resolve the apparent contradictions of economic growth and environmental improvement. That is, through a combination of clean technologies, institutional innovation and other smart practices, it is possible to reconcile the material benefits of industrial society with the satisfaction of environmental conservation: we can, indeed, have our cake and eat it too.

For example, across Europe the requirement for every domestic appliance to be marked with an eco-efficiency rating is designed to encourage consumers to make a change in purchasing practices toward "greener" products, so called because of efficiency improvements in their performance. Promoting the purchase of an "A-rated" refrigerator might be a small incremental change toward more sustainable behaviour, but it certainly does not challenge the rights of citizens to consume (and indeed may make negligible environmental improvements if, as the evidence suggests, consumers end up buying a bigger refrigerator which may draw as much power as their smaller old one). Ecological modernisation does not, therefore, address the problem of **consumerism**. On the other hand, for those who adopt such changes, it may offer a visible pathway of self-identifying with sustainable consumption and lead them toward more radical shifts in behaviour than simply purchasing "green" products (Hinton and Goodman 2010).

Food provides an interesting lens through which to view sustainable consumption, one that offers a means to go beyond narrowly eco-efficiency criteria in shopping choices (Goodman *et al.* 2010a). Food, unlike compact fluorescent light bulbs or hybrid cars, is "good to think, as well as to eat". That is to say, without too much effort, food has the power to convene many different perspectives, cross-cutting issues and questions: where does it come from? How was it produced and by whom? Under what conditions does that producer live, and do they share the same life opportunities as the person eating that food? Food, unlike any other product, entangles us in webs of relations that connect us to distant others: those who harvested the coffee berries, trimmed and packed those green beans, winched those nets on a stormy sea. Food should do this in a way that buying trainers does not, because of its essential materiality, it derives from ecological processes managed by other people however remote from us, and we ingest this material into our bodies: we are and become what we eat. Clearly, this should give us cause to think about what we put into our mouths and the grounds on which we choose between different foods.

The global agri-food system described in this book is one characterised by a drive to productivism, to scale, and to a belief in providing the consumer (at least in the North) with value for money. Providing a greater abundance of food at ever-lower cost has been the overriding feature of the post-war agri-food system. Northern consumers have come to enjoy this privileged position, provided with a cornucopia of products sourced from around the world. But, as we know, the externalities have been gradually mounting: the degraded soils, the depleted freshwater resources, a growing and precarious dependence upon fossil fuels. Meanwhile, the prices paid to farmers for their coffee beans, bananas and cacao continue to fall, so that the primary producers of our skinny lattes and our healthy or comforting snacks face further destitution. The problem is: how are consumer choices to be reshaped if evidence of degraded agro-ecosystems or environmental destruction or rural impoverishment is filtered out as the product moves through the supply chain? And even if such information were to be made available, would this lead to changed consumer behaviour?

This is an issue signalled by the term **distanciation**. Though lacking in elegance, this word suggests more than a high number of food miles separates primary producers from final consumers within the contemporary agri-food system; rather, there is also a lack of information, of knowledge, about the conditions of production and the supply chain through which those products pass. Hiding such information, making traceability difficult to establish, serves the interests of those who intermediate on behalf of consumers: the large processing, retailing and food service companies. Leaving consumers ignorant and lacking capacity to take responsibility for their purchasing decisions may be a satisfactory state of affairs for some, but not for hundreds of millions of others who desire to know more about where their food came from, how it was produced, and by whom. Of course, this also presupposes an ability to be able to engage with, and take action on, such information.

Consequently, thinking about our food cannot be only about its ecological consequences – as important as that is – but is also about its social, economic and, above all, *moral* dimensions (Goodman *et al.* 2010a). It is this ethical turn in thinking about food that has done so much to challenge the conventional food system, causing it to adapt and respond, as well as giving rise to what have been called **alternative food networks**, to which we return later.

Thinking ethically about food is regarded here as being synonymous with thinking sustainably. That is to say, it is aspirational and reflects a continuum of practices moving in the direction of improvement across a series of fronts (social, economic, environmental) rather than marking a definitive end point. Like sustainability, ethical food might be considered a rather fuzzy concept and may comprise a wide range of values and practices, depending on the subjective preferences of the consumer. In a recent paper, Tim Lang (2010) lists some ethical features of food as derived from a food industry research body survey of consumers in six European countries, and from an EU research project. Grouping these together gives rise to the following main categories:

● *Animal welfare* considerations, including products designated as free range, and those not tested on animals.
● *Fair trade*, representing the purchase from primary producers of products at a price that represents a living wage; labour standards, such as health and safety and guarantees against the use of child labour might also be considered here.
● *Sustainability considerations*, encompassing organic and other products that derive from careful husbandry of environmental resources, as well as efforts to reduce carbon footprints, packaging, food miles, etc.

- *Provenance*: the desire to source more local foods, to support primary producers in the region or country, or to defend particular speciality foods of cultural importance or gastronomic merit.

We can see that these represent a very diverse set of considerations. Some offer a relatively straightforward way for producers to lay claim to the label of ethical food (by improving the stocking density and conditions experienced by intensively reared livestock, for example), although this might have consequences for profit margins if such food products that then derive are still sold cheaply. Other considerations may involve more complex standards, more rigorous policing or the creation of new supply channels (organically certified or fairly traded produce). Clearly, to some degree all involve a combination of principles (the treatment of people, animals and the planet in a just and humane way) and information (involving transparency about methods of production and the dissemination of such details). Creating the conditions for an informed and educated public is vital, for only then can individuals choose to act ethically exercising personal responsibility. This means overcoming the "construction of ignorance" through which food commodities are divorced from the social relations of their production and provision, such that consumers know little about "the biographies and geographies of what we consume" (Cook and Crang 1996: 135). Arguably, fair trade produce serves to do this, by establishing through its label a connection between producers and consumers. Given the way in which it has grown and developed over the past twenty or so years, it is appropriate to explain its objectives and evaluate its success.

Fairtrade

The Fairtrade movement can be traced back to the 1940s and the marketing of handicrafts from the South in charity shops of the North, but it came of age with the creation of a certification system for coffee from the late 1980s. The Fairtrade Labelling Organization (FLO) was established in 1997 to act as an umbrella entity representing twenty or so certification initiatives around the world, almost 600 certified producer organisations, traders and over 1 million farmers. A Fairtrade label therefore appears on products that have been certified as meeting the required standards of the FLO. Principally, this involves primary producers organised into local cooperatives receiving a "fair" price for their product. This comprises a minimum baseline price, irrespective of what the prevailing international market price for the product might be, and then a premium payment that ensures it is above the market price. Further premiums are paid for produce that is certified organic or, in the case of coffee, which is "shade-grown" and consequently meets other ecological criteria. The purpose is to ensure that producers receive prices that cover their costs of production and that provide incentives for more sustainable farming methods.

Although these premium payments are directed, through the cooperative, to raising the income of individual farm families, additional measures are often available to provide a line of credit, to help support social development projects (installation of clean water, health schemes and so on), and to help strengthen long-term trading partnerships. This appears to provide a strong case for the claims that are made for Fairtrade, that it offers a vehicle for sustainable development and a means of lifting farm families around the world out of poverty. For consumers, this appears to be a win–win situation, particularly with significant improvements in the quality of the product. The logic, of course, is that by increasing consumption of Fairtrade products, the more benefit is derived by farmers: more really is good!

It is this logic that has seen Fairtrade go mainstream. Lobbying efforts in the USA have persuaded large corporations, including Starbucks, Dunkin' Donuts and McDonald's, to offer Fairtrade-certified coffee in its outlets. More recently, several global chocolate and confectionary manufacturers, including Cadbury's Dairy Milk, Nestlé's Kit Kat and Green & Black's, have converted to Fairtrade cocoa. For many Fairtrade proponents, this is demonstrable proof that if the product is good and the ethical case can be made, major food corporations will get behind it and provide the means to scale-up sales and revenue flows to farmers. For example, in the UK, retail sales of Fairtrade-labelled products topped £700 million in 2008 with over 4500 Fairtrade-certified products for sale in retail and catering outlets. Sales of Fairtrade cocoa products have been particularly strong, growing from £2.3 million in 1999 to £44 million in 2009 (Fairtrade Foundation 2009). This has resulted in significant material and social benefits accruing to the large cocoa-producer cooperatives in Ghana and the Dominican Republic and their members (Berlan and Dolan forthcoming).

Critics of the mainstreaming strategy argue, however, that having the right to use the Fairtrade label offers companies such as Nestlé, still the focus of boycott campaigns such as Baby Milk Action, a public relations opportunity to recover some ethical credit: the "halo effect"

(Holt-Giménez *et al.* 2007). Ultimately, Fairtrade-certified product such as coffee beans constitutes only a small proportion of the total volume of the purchases made by these large corporations, and the international coffee supply chain remains characterised by its "hour-glass" profile in which tens of thousands of coffee farmers and millions of coffee drinkers are linked through a small handful of international buyers and processors. With coffee prices remaining low, possibly not even covering costs of production, the Fairtrade premium of $0.10 per pound hardly represents a bonanza, and efforts to increase it have been resisted by certification bodies wedded to mainstreaming through conventional retailers. For critics of mainstreaming, such as Eric Holt-Giménez and colleagues, it is time for the Fairtrade movement to help small farmers build their market sovereignty, as they argue in Box 7.5.

Acting ethically remains a highly (though not exclusively) subjective process as each of us trades off some attributes over others – buying only organic may involve more food miles; buying a locally produced,

---

Box 7.5

Fairtrade and market sovereignty

Fairtrade certification in and of itself clearly does not change the dominant trade paradigm. So far, Fairtrade has not been able to advance any industry standards to rectify the imbalance of market power in the coffee market. In the "one size fits all" approach to Fairtrade, the bigger players capture the aggregated benefits of volume, while the smaller players cling precariously to smaller individual premiums.

It is unlikely that Wal-Mart, Starbucks, or Nestlé will advance a farmer-driven, movement agenda for social change within Fairtrade. They will attempt to sell as little Fairtrade coffee as possible at the lowest possible price, counting on their vast market power to keep Fairtrade farmers coming to them. This is not a reason to give up the Fairtrade market. On the contrary, it is up to alternative trade organisations, other development activists and consumers, to help poor coffee farmers grow not just their market, but their market power, not just their business, but their controlling share within the business.

Ultimately, the ability to establish higher, more equitable and integrated standards of fairness depends on the degree that the Fairtrade movement advances farmers' market sovereignty – the ability to produce, process, sell and distribute in ways that are fair and sustainable.

Source: adapted from Holt-Giménez *et al.* (2007), 13, 20.

out-of-season product may have a higher carbon footprint, and so on. Nevertheless, ethical considerations are now a feature of mainstream food shopping, as an IGD survey cited by Lang (2010) reports, with 69 per cent of those polled saying they actively looked for at least one ethical feature. This IGD report concludes that the market for ethical food has huge potential to grow further and is only held back by price and availability.

The modern food system, particularly as it exists in the North, has presented consumers with an unparalleled range of products, mostly all available year-round and possessing a blemish-free appearance, at prices where food purchases now account for a historically unprecedented minor share of household budgets. The ethical challenge comes not simply from choosing the "farm fresh" or "cruelty-free" labelled eggs or pork that are cost-competitive with similar unlabelled products and just as available at the local supermarket. It comes from a willingness to pay higher prices for foods routed through shorter supply chains; from actively searching out the local organic option with all its blemishes (and extra flavour) due to the lack of pesticides; or even choosing not to buy anything derived from animals at all. Indeed, the ethics of eating meat ultimately represents one of the most challenging individual decisions that we can take, although the grounds on which we might justify the decision can vary significantly. Given the environmental impacts of livestock rearing for meat and dairy production, as outlined in chapter four, it is as well that we explore the issue of meat as an aspect of sustainable consumption.

Reducing the consumption of meat

The ethics of eating meat has been widely debated over a very long period: there is a long tradition of vegetarianism in the East, although its practice in the West has been associated more with those who hold unconventional views and pursue alternative lifestyles. Nevertheless, there appears to be a gradual shift in public attitude that is permitting the right to eat meat cheaply to be more widely discussed. As Gussow (1994: 1113) noted, "many people believe that humans are morally obliged to stop behaving as if they were entitled to use animals as they please". The philosopher Peter Singer has done much to explore questions surrounding our attitudes to animals and it is illustrative here, while discussing sustainable consumption, to briefly rehearse some of the arguments in defence of eating meat. Singer and Mason (2006), though naturally highly critical of such reasons, set out several positions that might justify the maintenance of factory farming. These include arguments as follows:

- We have no duties to animals as they are incapable of having duties towards us.
- If animals have their own predator–prey hierarchies, then why cannot humans be entitled to form the apex of the pyramid?
- Meat, especially in the Western diet, is culturally and nutritionally significant and we could not sustain its low cost without factory farming.

Singer and Mason address these arguments by drawing parallels with our duties to others who cannot engage in contracts, such as children; our selective deployment of natural orders (there is nothing natural about our raising of animals); and through reminding us that societies that historically practised injustice and cruelty eventually took steps to rectify it, through the abolition of slavery, the emancipation of women, and measures to overcome racial prejudice. But while the ethical basis of factory farming might be easily exposed, what about free-range animals, where cruelty would seem forbidden? Singer and Mason make two main arguments in its defence.

- There is a co-evolutionary mutualism that binds certain species of animals to human societies, whether dogs and cats, or cows, pigs and chickens.
- Free-range animals live healthy, contented and fulfilled lives even if their years are prematurely cut short by the need for some of their body parts to occupy part of our plate.

The argument that these animals would not exist were there not demand for meat bypasses consideration of the experience of animals living their own lives in full possession of their intrinsic rights to exist. We invest enormous emotional value in family pets and are deeply distressed when their lives are ended prematurely. Why the same pain does not apply to lambs and piglets is due to the concealment of their slaughter. It is interesting in this regard to note that Singer and Mason draw attention to Michael Pollan's "lyrical portrayal" of Polyface Farm in Virginia, where six different food animals are raised, in what appears to be a fine example of an ecologically integrated sustainable agriculture. However, while the cattle, pigs, turkeys and sheep are trucked off to a slaughterhouse, a requirement of the US Department of Agriculture, the rabbits and chickens are killed on the farm. He describes his experience of participating in the process of killing and processing the birds in chapter twelve of his book, *The Omnivore's Dilemma* (Pollan 2006). The process can also be seen on the excellent documentary film *Food, Inc.* (2009). However "sustainable" the circumstances of their lives – and Singer and Mason note criticisms by others of the Polyface confinement system – the slaughtering of chickens there remains a fairly shocking sight. And this occurs on a farm widely admired for its humane practices and a willingness to be scrutinised. As they ask: how humane is humane enough to eat?

This reflection on the morality of rearing and killing livestock for the purpose of providing food to eat is not a digression: how we *think* about consumption really does matter. Building a sustainable agri-food system requires us to look closely at the ethical bases of our consumption decisions and, in this, there is a strong case for eating less meat. For example, we know that livestock consume around 40 per cent – and rising – of the world's grain harvests; and utilise, directly or indirectly, up to 80 per cent of the world's agricultural land. Yet they supply just 15 per cent of all calories (Stokstad 2010). Would eating less meat release more grain for those who are currently inadequately nourished? Would widespread vegetarianism dramatically improve the world food equation?

It is clearly the case that the rising appetite for meat around the world cannot be met without increasing the environmental burden on soils, water and the climate system. Yet, while factory farming in the North and middle-income countries has resulted in animals being taken off the land and put into cages, pens, stalls and feedlots for the rapid metabolisation of feed into food, in the rest of the world animals still perform multifunctional roles, converting material that humans cannot eat – grass, shrubs, crop residues and other wastes – into human food and providing many more services besides. A more sustainable agri-food system that included livestock – for meat, milk, eggs, wool, draft power, manure and other benefits – would probably comprise many fewer animals than at present. As there would no longer be a case for diverting such a large volume of cereals into the livestock feed chain, most of these would graze land either unsuitable for cultivation or in some rotational arrangement with annual crops. The grazing of many upland and grassland environments would continue in order to prevent those areas reverting to vegetation communities of more limited species diversity. As Gussow (1994: 1115) notes, without in any way promoting meat consumption, "it would be beneficial if everyone were to acknowledge the ecological appropriateness of omnivorousness".

Not choosing to become vegetarian, but living a life eating less meat, is going to become a powerful message in the months and years ahead. It is a message that is being broadcast by influential figures, including celebrity vegetarians such as Paul McCartney, and others of regard such as Dr Rajendra Pachauri, chair of the Intergovernmental Panel on Climate Change, both of whom spoke to the European Parliament in December 2009. Building on wartime messages to civilian populations to conserve food for troops serving overseas, the idea of "Meatless Monday" was resurrected in 2003 as a public health awareness programme in association with the Johns Hopkins Bloomberg School of Public Health, and was widely implemented across the Baltimore school system. In May 2009, the Belgian city of Ghent declared Thursday a meat- (and fish)-free day, while San Francisco city council made a similar ruling in April 2010. Meat-free days are a potentially powerful tool to promote a message of working toward more sustainable consumption, and we will be hearing of more city and regional authorities adopting them.

5. Reconnecting production and consumption for sustainability

It has become apparent over the past decade or so across Europe, North America and elsewhere that, particularly amongst people with higher disposable incomes, there is a generalised disenchantment with the modern food system. This may seem somewhat paradoxical, given its apparent success in delivering an unprecedented abundance and diversity of foods while consumers are spending the smallest proportion of their household budget on food. Yet the growing contradictions of productivism have given rise not only to the wide range of environmental problems that have been discussed in earlier chapters of this book, but also to a loss of trust and confidence in the food system. During the 1980s and 1990s, the agri-food industry was rocked by a succession of food safety crises. The most serious of these was the BSE (bovine spongiform encephalopathy) episode (colloquially known as "mad-cow disease"), which seems to have arisen from the use of animal feeds containing the rendered remains of other ruminants, including sheep. Once it became apparent that this disease was transmittable not only from sheep to cattle, but from cattle to humans, widespread anxiety ensued for a time in the UK and resulted in a severe impact upon the beef industry. A succession of other food scares in Europe (benzene contamination of wine, dioxin contamination of poultry feed, foot-and-mouth disease, salmonella infection in eggs) and the United States (*E. coli* in hamburgers and in spinach) had the effect of shaking consumer confidence for a time. While normal business was eventually restored following the introduction of wide-ranging food safety measures, and many consumers went back to shopping for and eating their preferred foods, such episodes helped to create the conditions for encouraging others to look more actively for traceability and other direct assurances about the quality of their food.

The emergence and development of a quasi-parallel system, widely referred to as alternative food networks or short food supply chains, was clearly given a boost by the experience of these food-safety failures, and evidence of its growth can be seen through, for example, the numbers of farmers' markets now in existence. The meaning of the term "alternative food networks" is discussed later, but first let us describe what they are. These networks comprise primary producers and food artisans; intermediaries such as market stallholders genuinely interested in promoting high-quality food; consumers and others (restauranteurs, food writers, social entrepreneurs) all of whom are engaged to some degree in building an alternative to the more standardised model of food supply. While they take on a variety of forms, what these networks have in common is an ability to redistribute value through the network so that money remains within the local economy, rather than to find it concentrated and repatriated back to the headquarters of corporate retailers. The high level of personal interaction that is a hallmark of exchange within these new networks offers a novel experience to customers who have become accustomed to the impersonal nature of human contact reduced to the barcode scanning of the contents of their shopping basket. Such experiences have the capacity to recover the collective human experience associated with the preparation and exchange of foods, and contribute to re-establishing the principles of trust, care, connectivity and health.

The emergence of alternative food networks has consequently arisen from consumers' concerns for human health and food safety, as well as other, wider ethical considerations (animal welfare, fair trade, sustainability). Such developments have demonstrated that food markets are not the result of some "invisible" hand external to the social world, but result from the active construction of networks by various actors in the food chain. This has encouraged the view that alternative food networks might help reverse long-established trends toward corporate concentration in the food sector and improve the share of total value accruing to primary producers. By breaking with the long, complex and logistically organised supply chains run for the benefit of the major corporations, food networks have the potential to forge new links between producers and consumers. Resocialising food – bringing it out of the highly individualised way in which consumers make personal choices within the range offered by supermarkets and other corporate suppliers, and more fully into the civic arena where matters of public good are given due weight and consideration, would ideally encourage a more reflexive and critical judgment about the relative desirability and quality of different food products, and offer the basis for more collective solutions to build local food systems for community development.

The term "alternative food networks" is not without shortcomings, but has proved a rather robust label by which to refer to that which is not the mainstream agri-food system. A distinction has been made between the way the term is used in Europe, where it has become closely associated with food quality, and in North America, where "alternative" signals oppositional or radical designs in reclaiming ownership of food supply chains (Goodman 2003). In fact, many local food networks represent a hybrid, comprising both more radical elements (the "old guard" of organic growing, say, who have steadfastly refused to compromise on their methods or philosophy) together with the new artisans who find their products on the shelves of a corporate multiple (and enjoy the increased volumes of sales that result). The word "network" is possibly a more accurate reflection of this reality and indicates the importance of thinking relationally about flows, processes and relationships (Kneafsey 2010). In short, there is an emergent property of alternative food networks that conveys a sense of flux, of dynamism and of *social* possibility. The conventional agri-food system, on the other hand, though dominant, resourceful and capable of continuous reinvention, especially in the way that it has sought to appropriate the language of alternative practices ("local", "organic", and with pictures of its specialist suppliers posted on the supermarket walls), remains monolithically corporate and still preoccupied by low cost (Jackson *et al.* 2007). It is indeed a case of David and Goliath.

One of the key strengths of alternative food networks is their strong association with territory: place, locality, the region. Indeed, alternative food networks frequently become synonymous with **relocalisation**, a term used to reflect the turn away from globalised food, characterised by uniformity and the result of highly mechanised process technologies operated in distant places. In place of these anonymous and standardised supply chains, alternative food networks offer a means to reconnect with the food, its producer and the place of production. Some of the following characteristics are associated with alternative food networks and the products that pass through them:

- *Food quality*: the embodied properties of the product – its taste, appearance and other sensual attributes – which give it distinction are accorded primary significance. The connection between quality and locality might become the basis of a regional cuisine, or for speciality products that might be considered worthy of designated origin labelling.
- *Tradition*: building on local food culture and typical products of the area. This might also involve utilising specific breeds of animal characteristic of the area, such as pigs that yield distinctive ham and other charcuterie. In general, novelty – the creation of entirely new products that do not claim at least some connection to a recognised food culture – is regarded suspiciously.
- *Craftsmanship*: the use of artisan techniques that include a high "hand-made" element involving skill and knowledge of the materials and technology utilised. Such skills are deployed, for example, in farmhouse cheesemaking, in adjusting for variations in milk quality (fat content) over the course of the year. Artisan-scale production invariably imposes a limit on total output volume.
- *Direct marketing*: short food supply chains provide the means to short-circuit the long, complex and rationally organised industrial food supply chains and offer a means for connecting directly with the consumer. They also enable more value to be retained locally. Short food supply chains cover a variety of initiatives, as illustrated in Box 7.6.

Even this short list offers a radically different basis for a food system. It is one marked by a potentially high degree of interaction between producers and consumers, which enhances trust and social capital that can spill over into other collective and community initiatives. It presents opportunities for capturing and retaining more economic value by primary producers, especially in peripheral rural areas, and might help to build synergy with other sectors, such as tourism. Above all, it restores the central importance of provenance as a vital ingredient of decent food. Although there are a number of ways to capture the emerging reconnection of production and consumption for sustainability (amongst them community-supported agriculture, box schemes, public

---

Box 7.6

A typology of short food supply chains

- *Face to face*: where the consumer purchases a product direct from the grower, farmer or producer. This category includes farmers' markets, farm shops and roadside stalls, pick-your-own, box schemes, home deliveries extending to mail order schemes involving telephone or online purchases. In all cases, authenticity and trust are mediated through direct personal interaction with the producer.
- *Spatial proximity*: although this short food supply chain is extended over longer distances, products are still retailed predominantly within the region of origin, invariably by people who have an understanding of the conditions of production, an appreciation of the attributes of the product, and therefore can guarantee its authenticity. While this category includes more complex institutional arrangements for sales, the regional identity of the product is at the forefront. Examples include consumer cooperatives, fairs, gastronomic tourism routes, arrangements between producer farm shops, regionally themed restaurant menus, and so on.
- *Spatially extended*: while distribution into national and even international markets may appear to contradict the notion of *short* supply chains, this category refers to products with a reputation and embedded with value-laden information about the place and characteristics of production. Regional specialities such as Parmigiano Reggiano cheese and Champagne wine, or Fairtrade products such as Guatemalan coffee are, in their different ways, products that are strongly differentiated from anonymous commodities, commanding a premium price and often retailed through distinctive supply networks. In order to maintain the "exclusivity" of the product, formalised codes, involving independent regulation and certification and established on a juridical basis, have been created to protect it from inferior imitations. Such codes include national and European designation of regional origin (PDO/PGI) or Fairtrade labeling (Parrott *et al.* 2002; Tregear 2003).

Source: adapted from Sage (2007), 151.

procurement initiatives, consumer co-ops), farmers' markets have proven one of the fastest-growing expressions of local food networks.

Farmers' markets

Around the world, markets are vital places that provide the backbone of the local food economy. Invariably occupying a strategic central location, markets draw together local people in the buying and selling of food. Markets have existed as sites of exchange from the earliest times, and were gradually formalised for the interests of taxation, regulating food quality and improving public health. More recently, markets have come under pressure from corporate retail chains offering "convenience and low price under one roof", while some town councils have been sufficiently backward-looking as to regard markets as a historical relic, to be swept away in favour of a shiny new superstore. Fortunately, markets are a tenacious institution, and market traders an especially stubborn group of people who refuse to relinquish their livelihood. And the arrival of farmers' markets, as a contemporary form of a long-standing tradition of rural people selling their produce, has had a major influence on food retailing across North America and large parts of Europe.

Farmers' markets appear to be an excellent way in which small growers, farmers and other food producers can market their produce direct to the public without the high transaction costs or minimum volume requirements associated with conventional food supply chains. Although they are of little relevance for producers of bulk commodities or those involved in contract sales to food manufacturers and retailers, they offer real opportunities for increasing returns to farmers. For example, Pretty (2002) compares the paltry 8–10 per cent share of each euro, dollar or pound spent by consumers on food that finds its way back to the farmer through normal marketing mechanisms with the 80–90 per cent when food is sold by that farmer directly to the consumer.

The power of local markets lies in their providing an explicit spatial alternative to conventional food supply chains. But this comprises more than shoppers making an effort to reduce the distances travelled by food. In recently conducted research, Milestad *et al.* (2010) discovered two kinds of learning experienced by both farmers and shoppers in two Swedish farmers' markets. The first, instrumental learning, comprised information about factual matters (prices, supply, production conditions, etc.) exchanged during face-to-face interactions. To this one might add the opportunity for shoppers to find out more about reconnecting with their regional "foodshed". The second, communicative learning, concerns the establishment of shared norms and values and the way these enhance trust and the ability of each party to help fulfil the needs of the other (farmers endeavour to produce the best food they can at a fair price; customers turn up every week to buy it and provide feedback). Critically, both provide the means to enhance learning through fostering greater ecological knowledge (what can be grown locally and when) as well as better appreciation of the nature of the food products. For Milestad *et al.*, this offers the opportunity to enhance the *adaptive capacity* of both farmers and shoppers at farmers' markets and therefore the resilience of local food systems. Building upon the communication and trust that exists between them, shared interpretations about food and farming conditions will inform the agro-ecosystem managers (farmers) to make sustainable decisions for the benefit of both.

Consequently, farmers' markets provide the basis for transactions, but also dialogue between producers and consumers, with the potential for recovering a moral economy around food. They hold the potential for significantly challenging conventional production, retail and consumption patterns by setting out alternatives that embrace ecological citizenship. Moreover, evidence from the United States has demonstrated that markets do have nutritional and health benefits, especially when located in low-income areas, by making available fresh, affordable food.

6. Planning and creating sustainable food systems

The previous section makes a strong case for recovering a sense of territory and place in the development of alternative food networks. Considering localities as sites where food can be considered ecologically, culturally and socially embedded, and where producers and consumers can meet to engage in communication for social learning, are vital first steps in working toward a more sustainable food system. Yet there has been a strong reaction by some scholars to the notion of local food systems, arguing that there is nothing inherently superior – ecologically, in terms of social justice, nutritionally, or with regard to food security – about the local (Born and Purcell 2006). There is some truth in this, and it reminds us to be aware of a priori assumptions that sustainability can be achieved only at the local level. Unfortunately, however, efforts to chart local food initiatives are often met with accusations that they bolster the status quo, reproduce existing social privileges and fail to achieve social equity: they are therefore fundamentally elitist. Such a charge partly reflects differences in perspective between North American and European scholars, noted earlier in relation to alternative food networks: the former bring to bear a more radical social justice approach in their expectations for local food. Yet such criticism is itself guilty of a priori assumptions, that rescaling food activities to the local level cannot provide a solution to social justice: this must surely be determined on a case-by-case basis.

While acknowledging that relationships of proximity do not guarantee social equity, the belief here is that prospects for securing greater sustainability in the food system are as likely to derive from initiatives generated "from below" as from top-down policy. In fact, both are likely to meet somewhere in the middle, where the city and its hinterland, or a rural region and its network of market towns, may become the crucible for fashioning new, sustainable and, hopefully, more equitable food systems. The key to this process is going to be the willingness and capacity of nation states to engage in such endeavours within the public realm. Unfortunately, all around the world, though with some notable exceptions, the paradigm of neoliberalism has come to dominate. Briefly, the neoliberal model holds that the market is the most effective and efficient instrument for the allocation of goods and services and should, at all costs, be allowed to operate in an untrammelled and unrestricted fashion. The dominating discourse of neoliberalism over the past thirty years has not only caused the state to retreat from areas that were otherwise regarded as a necessary part of its responsibilities, but has also left it entangled in bureaucratic procedures where market instruments, such as value for money, have become institutionalised. This has held back the state's capacity to help bring into the mainstream a range of more sustainable goods and services, through the agency of public procurement, at least until recently. This is discussed in more detail later.

The retreat of the state has been mirrored by the rise of powerful corporate interests that have come to provide consumers around the world with their needs and wants. In the realm of food, we have noted how so much variety is available at historically unprecedentedly low consumer prices, yet at significant external cost to the environment. But there is also the cost to human health and wellbeing. In the affluent countries of the North, as well as in growing numbers of middle-income countries, evidence points to a serious and mounting burden of diet-related ill health. Numbers of obese and overweight now exceed those who are hungry and malnourished. The medical evidence clearly demonstrates that this contributes to rising incidence of diabetes, heart disease and cancer, suggesting that state health services around the world will face a significant funding challenge into the future. Yet the nutritional evidence is also very clear: if we were to replace diets high in salt, animal products, oils and fats, processed sugars and carbohydrates, with diets of greater variety containing more fresh fruit and vegetables, health indicators would improve. This is not going to happen, however, through the "nanny state" telling us to eat up our greens!

Planning for healthier, more sustainable and more socially just food systems is in its infancy, but is in a position to grow quickly. It may be surprising that it has remained dormant for so long, given that food policy played such a vital role during the Second World War and in earlier upheavals. Though it took different forms in different countries, the rationing of food was widely practised and revealed enormous health advantages from such interventions; encouraging people to grow their own food ("dig for victory") was also regarded as a necessary and patriotic act. The experience of wartime was a major post-war stimulus to productionism in Europe and North America, but eventually this top-down approach to food policy as health and welfare began to fragment and fade (Lang *et al.* 2009). This now looks set to change, as food planning becomes a rapidly growing and vital arm of public policy, given the emergence of a number of complex developments, including:

● the food price surge of 2007–08 that has caused many countries to reconsider the basis of their food security;
● climate change effects, including water scarcity and heat stress, with the experience of Russia during the summer 2010 especially striking;
● the escalation of land-leasing schemes, where China, South Korea and Saudi Arabia, amongst others, are securing access to hundreds of thousand of hectares of land in Africa and Asia;
● rapid urbanisation, raising questions of how cities will feed themselves – at times of rising food prices and shortages, it is urban areas where protests are greatest (Morgan and Sonnino 2010).

Food planning is not confined to professional and practising urban and regional planners – indeed, the American Planning Association took some persuading to agree that food was even a legitimate focus of attention until the publication of its Policy Guide in 2007 (Pothukuchi 2009). Rather, food planning involves bringing together a diverse range of people concerned with public health, social justice and ecological functioning: the essential elements for creating harmonious, healthy, creative and sustainable cities and regions. As food planning takes shape in many different places around the world, there are a multitude of examples of good practice and opportunities for learning.

For example, Toronto has had an enormously successful Food Policy Council in place since 1991, and this provides a platform for a wide range of representatives to devise ways of making the city's food system more sustainable. London, despite changes in political administration, appears to have maintained a commitment to its food strategy designed to strengthen its food security, and this includes increasing production in and around the capital (Reynolds 2009). In New York City, where obesity and overweight affect more than half the adult population and 43 per cent of schoolchildren, a number of healthy-eating strategies have been introduced to regulate junk food, promote the consumption of fresh fruit and vegetables, and support food-growing initiatives (Morgan and Sonnino 2010), including the development of rooftop agricultural greenhouses. A striking feature of these and other world city food strategies is the importance being attached to urban primary food production.

Urban agriculture

With the majority of the world's population now living in cities, and with this proportion set to increase, urgent attention is now being given to how these places will be provisioned. Whereas urban centres once grew in line with the capacity of their rural hinterlands to provision them, cities that became politically powerful and economically successful could extend their supply routes further afield, ultimately reaching anywhere in the world. As transport infrastructure networks multiplied, supported by cheap oil, cities were no longer entirely anchored to national territory and could grow dramatically in size. Yet limits now appear to be reached: how will such places be provisioned with food and freshwater in an era of peak oil and climate change? Given the creaking infrastructure in many cities (including those regarded as wealthy), how will they cope with solid and liquid wastes besides the need to reduce atmospheric emissions? And, critically, how will such cities improve the health, diet, food security and livelihoods of the urban poor, most of whom make up the ranks of those clamouring for greater social justice?

These are huge questions requiring complex and multilayered responses that will need to be tailored to individual cities. But one strategy that is being widely considered and actively developed is urban agriculture: the growing of primary food within and on the edge of the city. Expertise in this field is widely distributed, and much of it can be found in many cities of the South, where urban gardening in backyards, on vacant land along roads and railway lines, even roofs, can be found. Of course, there is a strong tradition of urban allotments in many cities of the North, and recent years have witnessed a revitalisation of these spaces after years of municipal neglect. There are many advantages to supporting the growth of urban agriculture:

● Improving urban food security and nutrition: households that grow some part of their food needs are likely to have a more diverse diet, with better nutritional outcomes.

● Poverty alleviation: eating food from their own plots reduces household expenditure on purchased foods, while for those that produce a surplus there is the opportunity to barter or sell food to neighbours. For peri-urban farmers occupying a little more land, production of fruit and vegetables for local markets is a commercial reality. Such ventures also help create paid employment.

● Improving the urban environment: urban agriculture can make productive use of land that is otherwise unfit for construction (situated under power lines, in flood-prone areas, along embankments). Not only does it help to "green" the city in terms of aesthetic appearance, it also helps reduce the urban heat-island effect of higher urban temperatures (e.g. multiple benefits of fruit and nut trees for shading, sequestration of CO_2, increased evapotranspiration). Of course, every tonne of food produced within the city reduces that which must be brought from further afield, with the likely result of saving energy and carbon emissions.

● Recovery of waste streams: it was once common for human bodily waste ("nightsoil") to be collected and used on the land to fertilise crops; this was once regarded as the reason for the success of Chinese peasant agriculture. While precautions need to be taken to prevent the spread of pathogens, composted organic wastes including excrement can be used as a source of soil fertility. Wastewater can also be used to irrigate plots, gardens and fields, reducing discharge into streams, rivers and other surface-water bodies.

● Social inclusion: the greater proportion of urban farmers are women, particularly in Africa, and there are significant benefits to be gained from helping them develop these activities. Small-scale food growing can fit well with other commitments women may have (childcare, informal wage-earning), requires little by way of start-up costs, and delivers benefits (fresh food, cash income) directly to family dependents. Women are also frequently at the heart of the community so that such work can move beyond individual households and help to create wider synergies, involving those who might otherwise be unable to work (those with disabilities or long-term illness such as AIDS) and help to build social capital. It has also proven important for immigrants, particularly those arriving from countries of the South into cities of the North, as they can put their horticultural skills and knowledge to use growing food plants of cultural and culinary significance.

While not promoting urban agriculture as a panacea for all social ills, some of the features noted above do offer some potential for enhancing the quality of life in cities. Although all might enjoy the recreational and therapeutic aspect of such work, it has the potential to be more than a middle-class leisure activity (the hobby allotment) in cities of the North, and genuinely to offer a means to improve the diet, health and wellbeing of all, including those in low-income and disadvantaged neighbourhoods. Bringing people together, either to work side-by-side in small, individually controlled plots, or collectively in community gardens, creates opportunities to engage in reflection on the injustices of the contemporary agri-food system and to address our alienation from nature. For McClintock (2010), urban agriculture has the potential to help re-establish a relationship between humans and our biophysical environment by reintegrating intellectual and manual labour. If we can recover a sense of the link between ecological and human health, we would cease to be so complacent about the damage inflicted on nature in the name of producing our cheap food. Urban agriculture opens a space around which food planners might locate opportunities to challenge the agri-food system by offering a more equitable, healthy and ecologically sustainable alternative.

Public procurement of food

A second dimension of food planning, deserving of attention here as vital to the creation of sustainable food systems, concerns the role of public procurement. This essentially refers to the role of the state as customer; with an enormous budget at its disposal, it makes decisions about the purchase of goods and services for public provision. In relation to the UK, Morgan (2008) traces the lamentable performance of successive governments in spending public money on defence, civil engineering and large digital infrastructure contracts, too many of which were significantly delayed, suffered huge cost over-runs, and were ultimately irrelevant (e.g. the Millennium Dome). He interrogates the reasons that might explain this pattern of failure, and notes the prevalence of a bureaucratic culture extolling design over delivery; the lack of integrative thinking among and between civil service departments; and a failure of confidence to assert public sector priorities over private sector interests. Clearly, the institutionalisation of neoliberal thinking after 1979 did mean that market-based instruments such as value for money came to crowd out other criteria in purchase decision-making. Nowhere has this been more shocking than in the public provision of food.

Public canteens are where the most vulnerable consumers are found: in schools, hospitals, care homes and prisons. Yet it is widely acknowledged that the quality of food in such institutions has not been of the highest standard. Part of the problem for providers (local authorities and related agencies rather than central government) has been the bewildering array of national and international regulations prioritising issues of food safety and economic competitiveness over quality considerations or sustainability (Sonnino 2009). In the UK, where public sector spending on food and catering services amounts to around £1.8 billion per year, the provision of these services has been dominated by large multinational companies that supply pre-prepared and standardised food on the basis of best value. Such food results in the deskilling of catering staff and offers scope for economies of scale, giving large firms a cost advantage and the opportunity to distribute food over large areas. This excludes smaller local producers and suppliers, and food quality and provenance disappear along with sustainability (Kirwan and Foster 2007).

Reforms in the UK began to emerge from 2002, following the publication of the Policy Commission on the Future of Farming and Food – which pointed to the role of public bodies in helping to create a more sustainable food economy – and a report commissioned by the devolved Scottish Parliament on redesigning a school meals strategy. A television series fronted by celebrity chef Jamie Oliver in 2005 further helped fuel an appetite for change. The evidence also suggests that regional initiatives for improving the quality of school and hospital food are often driven by energetic and innovative individuals willing to work outside their immediate functional role in the pursuit of a wider vision. There are now emerging a range of interesting and successful projects that are having a real impact on the quality of meals served, with multipliers for local economies and the wider environment. They demonstrate the power of effective public procurement for the creation of sustainable food systems. In the space that remains, just one example is used to illustrate this potential.

In the County of East Ayrshire in West Central Scotland, a very successful public procurement project has run with the goal of improving the quality of food served to primary schoolchildren. In examining this project, Roberta Sonnino (2010) points out to the critics of local food initiatives that, far from operating in a safe space for the privileged, the area is marked by social, economic and health indicators that are below the Scottish average (which itself is close to the bottom of European league tables on such measures). Consequently, the success of this school meals programme highlights the potential for improving diet and health in a context where levels of obesity in school children have been on the rise for the preceding five years. What the study also reveals is the enormous capacity for public agencies to play a leading role in setting sustainability objectives and to work in a supportive and facilitating way with local food producers in order to spread economic benefits within the region. The study is summarised in Box 7.7.

Growing sustainability from the roots?

Urban agriculture and public procurement have been highlighted here as two instruments amongst many through which public policy can play a significant role in creating more sustainable food systems. Clearly, their effectiveness depends to a significant extent on the vision and political will of those in positions of power in regional and city government. However, working towards the creation of a local and more sustainable food systems does not require the indispensable support of local authorities: much can, and must, be done by groups of individuals who are active in their community. This is vital for initiating change, mobilising others around alternative visions and providing a source of pressure on local politicians – who will invariably engage with a process that appears to be popular, successful and not hugely demanding of financial resources. Of course, local people will mobilise around different concerns, depending upon the circumstances: some will be engaged in anti-hunger/anti-poverty/social justice initiatives; others will be more explicitly environmental, concerned with the prospects of peak oil and climate change and what this might mean for food security; while others will seek to protest a deterioration in food quality and want to recover and defend local culinary tradition, distinctive regional products and agriculture. Irrespective of their motives, all these initiatives share a sense of active civic engagement, and the performance of *citizenship*. Consequently, in the creation of these new associational arrangements through which to recover a sense of trust and morality in the food

Box 7.7

School food in East Ayrshire

In August 2004, East Ayrshire council began to trial fresh organic and local food in one primary school. The feedback from children, parents and staff was so positive that, a year later, the pilot was extended to another ten schools. Council staff devised a tendering process that helped to involve small and organic suppliers in the school meal system. This included the following strategies:

- the strict "straightness" guidelines for vegetables (recall chapter five) were made more flexible;
- to enable small local suppliers to compete with larger companies, the tendering contract was divided into nine product categories;
- to improve the quality of the supplied foods, equal weight was attached to price *and* quality. Quality considerations included:
  - an ability to supply to deadlines
  - the capacity to provide Fairtrade, seasonal and traditional products and to meet ethnic, cultural and religious diet needs
  - food handling arrangements and facilities, to include safety considerations, but also training opportunities for staff
  - use of resources: to reward suppliers' contributions to biodiversity, minimising packaging and waste, and compliance with animal welfare standards.

As a result of this innovative procurement approach, East Ayrshire has significantly upgraded the quality of food served to children. Today, 50 per cent of the ingredients are organic; 70 per cent are locally sourced; and more than 90 per cent of all food is unprocessed. As for the cost of the meal, local suppliers offered prices that were, on average, 75 per cent higher than those offered by larger national firms. But the effects within the schools came down to just an additional 10 pence per meal, bringing the cost to £1.52 per meal in 2007. Larger portion sizes and higher nutritional quality were subsidised by the Council. Environmentally, the reforms produced important outcomes: food miles were cut by 70 per cent, schools were left with less packaging waste, and the involvement of local suppliers had a significant multiplier effect in the local economy. A survey conducted in 2006 revealed that 67 per cent of children thought that school meals tasted better, and 77 per cent of parents believed the scheme was a good use of the Council's money. School staff felt that the local sourcing of produce used in the meals created positive links among schools, communities and the environment. In 2008, the Council decided to extend the reform to forty-two schools, leading to the investment of £250,000 in the local economy.

Source: adapted from Sonnino (2010).

system, individuals can no longer simply be regarded as "consumers", which has a rather passive connotation, but as *food citizens*.

There are a multitude of local projects around the world that illustrate the extraordinary development of this new social movement around food. Although the United States has a diversity of initiatives working to advance the cause of a more sustainable food system, one of the first issues to focus citizen-consumer actions was hunger and food poverty, and ways of working collectively toward its elimination. The Community Food Security Coalition brings together over 300 organisations across North America in its efforts to:

> build strong, sustainable, local and regional food systems that ensure access to affordable, nutritious, and culturally appropriate food for all people at all times. We seek to develop self-reliance among all communities in obtaining their food and to create a system of growing, manufacturing, processing, making available, and selling food that is regionally based and grounded in the principles of justice, democracy, and sustainability.
>
> (www.foodsecurity.org)

Its work in Hartford, Connecticut, one of the first projects to emerge within this paradigm, is described in Box 7.8.

In Europe, too, there are many diverse models of collective and associative efforts around either the purchase or production of food. A successful illustration of the former can be found in Italy, which has witnessed the rapid spread of solidarity-based purchase groups (Gruppi di Acquisto Solidale, GAS) since they were first formed in 1994. The principal motivation for the formation of GAS is their stated opposition to the prevailing model of consumerism. In other words, they seek to create examples of more sustainable consumption and do this as follows:

> When a purchasing group doesn't search for just for the cheapest price, but instead puts people and environment before profit, the group becomes a solidarity purchasing group. A solidarity purchasing group chooses the products and producers on the basis of respect for the environment and the solidarity between the members of the group, the traders and the producers. Specifically, these guidelines lead to the choice of local products (in order to minimise the environmental impact of the transport), fair-trade goods (in order to respect disadvantaged producers by promoting their human rights, in particular women's, children's and indigenous people's) and reusable or eco-compatible goods (to promote a sustainable lifestyle).
>
> (Gruppi di Acquisto Solidale 2010)

Box 7.8

Community food security in action

Hartford is Connecticut's capital city, with a population of 120,000. Almost 31 per cent of the residents live below the poverty level, making Hartford the second poorest city in the USA. Under the leadership of the Hartford Food System (HFS), a private, non-profit organisation formed in 1978, the city has developed numerous food projects and public policy initiatives. These include a Second Harvest food bank, farmers' markets, community gardens, a community-supported agriculture farm, new supermarkets (the city has lost eleven supermarkets since the early 1970s), new bus routes that connect low-income neighbourhood residents to suburban supermarkets, a state anti-hunger coalition, farm-to-school programmes, farmland protection programmes, diet and health initiatives (city residents suffer from obesity and diabetes rates twice as high as the state average), improvements in federal food assistance programmes, a City of Hartford Food Policy Commission, and a State of Connecticut Food Policy Council.

HFS took a system approach from the very beginning, because it saw that the food problems facing city residents were complex, deep, and often interrelated. For instance, small and medium-sized farmers in the region were going out of business because they could not get a good price for their products. Low-income city residents could not find affordable retail outlets where quality, fresh food could be purchased. HFS opened the state's first farmers' market in 1978 in downtown Hartford, giving farmers a retail outlet and consumers an affordable and accessible place to purchase healthy food. To encourage farmers to sell in low-income neighbourhoods across the state, HFS combined on-the-ground project-organising work with a policy initiative (an approach that it would often repeat) that established the Farmers' Market Nutrition Program. The Program provides federal and state funding for vouchers that low-income mothers and senior citizens use to purchase produce at farmers' markets. Today, there are sixty-five farmers' markets across the state, serving nearly 200 Connecticut farmers and almost 60,000 low-income individuals (and hundreds of thousands of non-low-income residents).

HFS took the lead in establishing the City of Hartford Food Policy Commission, which comprises representatives from government and non-government sectors. Not only did the Commission let the general public know that there were programme deficiencies adversely affecting the food security of city residents, they let the City Manager, Mayor and City Council know as well. They convinced the appropriate city authorities and agencies to accept responsibility for the problems and engaged them in processes to find solutions. As a result of these processes, and the subsequent improvements in programme delivery, the quality of, and participation in, all these programmes increased significantly.

Source: adapted from Winne (n.d.), 6–7.

A census conducted by the network indicated that there are more than 600 GAS groups operating in Italy today, connecting about 75,000 families. Not all GAS groups deal exclusively with food: some are also involved in negotiating collective purchase of energy, telephone and other consumer needs. Another feature is that each group's membership interprets the principles in a way that suits its particular and local needs: they are self-organised and voluntary, with room to develop very close working relationships with participating farmers and other producers. Such is the concern on the part of these citizen-consumers to support producers that another network initiative has emerged to strengthen the capacity of GAS groups to manage the logistics of direct purchasing and distribution (*Reti di economia solidale*, RES). Meanwhile districts of solidarity-based economy (DES) try to link, support and even fund producers in order to integrate them into territorially based, directly managed short food supply chains (C. Grasseni, personal communication, 2009).

A recent phenomenon in the anglophone world has been the rise of the Transition Network, one of the highest-profile of the many relocalisation movements that have emerged during the past two decades. Founded in Totnes, Devon in 2005, one of the key triggers of its creation is an appreciation of the inevitability of peak oil (as discussed in chapter four) and the need to prepare for its eventual arrival. The Transition initiative brings together an intriguing blend of disparate ideas and influences: the principles of permaculture as a means of establishing self-sufficient, diversified and permanent cultivation of food; aspects of sustainability thinking, especially the building of resilience; and an array of behavioural change instruments to facilitate personal and community awareness raising, motivation and growth (Bailey *et al.* 2010). The promotion of the Transition message has successfully resulted in the almost viral spread of relocalisation initiatives. At the time of writing there were 636 Transition initiatives listed on the Transition Network website, 320 of which are officially designated Transition Communities, 170 of these in the UK, and almost all of the rest in English-speaking countries. The other 306 communities are in the process of mulling over the prospect of applying for Transition status, and this includes "mullers" in other European countries.

Although it addresses a range of concerns that follow from the realisation of peak oil, such as the need for communities to prepare an "energy-descent plan", the Transition movement places considerable importance on the growing of food. This provides a practical expression of reconnecting with the earth, an anchor into the locality, an opportunity to exercise a degree of autonomy from the global food system, and the basis to connect with others in the creation of community. The multiplicity of ways in which food growing can provide a focus of activity is striking; some examples are noted in Box 7.9.

One of the real strengths of the Transition initiative is that it has taken peak oil and future hydrocarbon scarcity, and used these to create a practical and empowering message for those who no longer wish to remain passively engaged in the global agri-food system. It speaks to a

---

Box 7.9

Growing local food

In back gardens (and even front yards), growing a few vegetables for household consumption need not be done in isolation, but can be part of a community effort to share gardening knowledge, tools, seeds and seasonal gluts. Sometimes a bit of helpful advice is all that some people need to get growing. And for those who have gardens, but not the time or labour power to work them, and for those who are keen to grow food but lack land, there is the garden-share scheme, whereby garden owners and those who want to garden both agree access to a site, with a share of the harvest being provided to the owners.

Community gardens and orchards exist in a variety of forms and scales around the world, but have in common that they are run and maintained by members of the community. Gardens might be established on whatever land is available – a derelict lot, abandoned brownfield sites, neglected and overgrown "green" space. Where owners of the land are not consulted prior to planting – known as "guerrilla gardening" – such efforts can be used as high-profile events to put pressure on local councils to make land securely available to the community.

Community-supported agriculture is becoming an increasingly important part of local food networks, especially in the USA. In essence, this comprises farms in which community members become shareholders and contribute to making decisions about what is to be grown. For farmers, this represents a radical change to the way in which their farms were previously run. Now, in place of producing a limited range of commodities for distant markets with uncertain prices, they produce a wider range for the local community at a price agreed at the start of the season. This serves to share risk, build trust and bridge the producer–consumer divide, and contributes to strengthening regional food security.

Source: adapted from Pinkerton and Hopkins (2009).

responsibility for making food security a local, even a personal, issue. Moreover, as the medical evidence makes absolutely clear, we need to be eating more fresh fruit and vegetables and taking more exercise: can growing some of one's own food become a means to facilitate popular engagement with the creation of a more sustainable and healthier food system?

Of course, as noted earlier, the emphasis on relocalisation does not mean that issues of greater equity and social justice are addressed: a major source of criticism is that the steering groups running Transition communities are dominated by the well educated middle class (Bailey *et al.* 2010). Nevertheless, it does help to stimulate thinking about the design of a future food system in which principles of resilience, sustainability and food security for all are likely to require a greater degree of social innovation than "top-down" messages about greening the contents of our shopping basket. In this respect, Transition and other relocalisation initiatives have a vital role to play in working with local government bodies and other civic groups to help plan and create more sustainable food systems.

7. Summary

This has been a wide-ranging review of sustainability, tracing some of its underlying principles, and applying it to the realms of production and consumption. It ought to have made clear that sustainability is fluid and relational, contested and complex, and above all locally specific. It is a term that is said to be socially constructed, meaning it can be used by different people, in different ways, to represent different things. However, that does not make it worthless, but rather demands that we make explicit what it is we wish to sustain. As Leach *et al.* (2010) observe, sustainability must refer to explicit qualities of human wellbeing, economic justice and ecological integrity: it must always embody environment, economy and society. Efforts to bracket off "sustainable" solely in relation to the economy ("sustainable growth") or the environment usually reflect its capture by powerful interests for their own ends, and should be resisted. In this regard, Maxey's observations are entirely pertinent when he refers to the growing number of claims for "sustainable" practices in relation to food provisioning, when there are so few for those that are "unsustainable". It may be the case that encouraging a more explicit rendering of the values, practices, scale and impacts of truly sustainable food production will expose the shortcomings, concealment, profit-seeking motives and unsustainable dynamics of the mainstream agri-food system (Maxey 2007).

Further reading

Leach, M., Scoones, I., Stirling, A. (2010) *Dynamic Sustainabilities: Technology, Environment, Social Justice*. London: Earthscan.
This book builds on a body of work conducted at the University of Sussex over the past two decades or more, outlining the tools that are needed for thinking about a diversity of pathways to sustainability. It is global in scope and explores other sectors and issues besides food.

Pretty, J. (ed.) (2005) *The Earthscan Reader in Sustainable Agriculture*. London: Earthscan.
Twenty-seven chapters comprising a variety of different perspectives: North and South, social and agroecological.

Lang, T., Barling, D., Caraher, M. (2009) *Food Policy: Integrating Health, Environment and Society*. Oxford: Oxford University Press.
The authors outline an integrative approach to food policy, which they call "ecological public health". Although dealing very largely with the UK, it offers a thoroughly comprehensive analysis of why sustainability should underpin food planning.

APA (2007) *Policy Guide on Community and Regional Food Planning*: Chicago, IL/Washington, DC: American Planning Association, www.planning.org/policyguides/adopted/food.htm
An extremely valuable resource for anyone remotely interested in food planning.

Maye, D., Holloway, L., Kneafsey, M. (eds) (2007) *Alternative Food Geographies: Representation and Practice*. Oxford: Elsevier.
This book provides a state of the art on thinking about alternative food networks, although the field is changing quickly.

Pinkerton, T., Hopkins, R. (2009) *Local Food: How to Make it Happen in your Community*. Totnes, UK: Green Books.
There are some points in this book that reveal an often idealised and inadequate grasp of the wider issues surrounding food. But as a "how-to" handbook it will probably become indispensable.

And on the web

Meatless Monday: www.meatlessmonday.com/baltimore-schools

The Transition Network: www.transitionnetwork.org

The Fife Diet: www.fifediet.co.uk – central California this is not! Brings a new dimension to the notion of locavorism.

Making Local Food Work: www.makinglocalfoodwork.co.uk

8 Conclusion

This book set out to explore the relationship between the environment and the food that we eat. The environment is represented as comprising a set of resource stocks (soils, water, biodiversity) and waste sinks (the atmosphere, water bodies) that are that are sustained by a diverse range of ecosystem services (provisioning, regulating, supporting). A variety of evidence is presented to support the argument that the contemporary global agri-food system imposes demands on the environment that we might regard as unsustainable. In other words, that the rate of depletion, degradation and loading threatens the long-term capacity and functional capability of these resource-provisioning, waste-sequestering and ecosystem services.

With regard to food, it is recognised that there is little space in the book to engage with the materiality of the things that we eat, or the significance (cultural, psychological) that we attach to them. A broad distinction is made between primary and final foods, the former comprising those products in their "natural" state (including raw meat and fish) that are either consumed directly or become ingredients for processing and manufacturing into final foods. This category accounts for the greater part of the agri-food industry, where cheap raw materials are transformed and valorised into branded products for international, if not global, promotion. That many of these products offer little nutritional value, are effectively "empty calories" (such as soft drinks), but deliver large profits to corporate stakeholders demonstrates that the agri-food system was never designed to achieve nutritional security for all. Indeed, that such products appear to be directly associated with rising levels of adiposity and diet-related ill health suggests there is a need for greater scrutiny and public policy regulation in practices designed to encourage increased consumption, for example by children.

The global agri-food system has grown and changed at a remarkable pace: having transformed food consumption practices in the affluent countries of the North since 1945, it is now engaging in a similar undertaking in middle-income countries of Asia and Latin America and prosperous locations elsewhere. New production systems are being established to help achieve the large-scale, low-cost model in these new emerging markets. Intensive livestock rearing is a particular feature as demand for meat rises, with consequences that spread far beyond the factory farms. The agri-food system has recruited millions of farmers and agricultural workers across the world in forms of contract production for exotic and novel primary foods for supermarket shelves. Its power and reach are truly staggering.

From an ecological perspective, the agri-food system is a major contributor to climate change, loss of biological diversity – including a narrowing in the genetic base of our food crops – as well as diminishing the stock of soil and water resources. From a social perspective, the system is guilty of production practices that are so focused upon lowering costs that it resorts to the cheapest, most vulnerable labour available – migrant and often illegal – to perform harvesting from fields, poly-tunnels and shellfish beds. Moreover, its promotion of cheap food has done much to exacerbate issues of poverty and social inequality, as discussed in chapter six. Finally, and from an economic perspective, it is apparent that reliance upon markets as the sole mechanism for the distribution of food, which depends upon an ability to pay to express demand, is responsible for almost one-fifth of the world's population experiencing hunger. Yet the system is clearly engaged in a process designed to extend the commoditisation of our food supply by producing new, "healthier" products designed to overcome the flaws and shortcomings inherent in earlier versions. "Low fat", "no sugar" and "less salt" labels offer manufacturers the opportunity to sell more products, encouraging consumers to indulge, seemingly with no cost to waistline or cardiac health. The rise of nutriceuticals, or functional foods such as probiotics, promises to take food further into medicine – provided that we can pay (Wilkinson 2002).

Given this, it is evident that we need a different model through which to procure our food. It is neither possible nor desirable to set out a map or blueprint of how this will be constructed, but it should surely begin by looking closely at some of the least desirable features of the current system, and then work toward achieving some consensus for their elimination. And there are some fairly obvious candidates. The purchase of single-use, small plastic bottles of water to slake our thirst; the expectation that we should eat meat products at every meal; throwing away unopened food because of "best before" dates: these would be amongst my own candidates for immediate attention. Then there are opportunities to shop differently: to buy Fairtrade-labelled products rather than the global brands promoted by Hollywood actors; to customise local shops rather than corporate retailers; to attend a farmers' market and make a connection with a local producer.

As this process proceeds – arranging for the direct purchase of food from a producer (e.g. a box scheme), participating in a community-supported agriculture scheme, or beginning to produce some part of one's own food directly – the issue of sustainability comes more sharply into view. Sustainability has been presented here as a contested concept comprising a set of important principles and system properties; it has also been presented as a normative model and a direction in which we should move, though this would be subject to considerable local negotiation. Like Maxey (2007), I believe sustainability offers a more useful and robust way of labelling this different approach to securing our food than the use of "alternative food networks". A sustainable food network really does offer the potential to create a viable parallel to the prevailing agri-food system, recognising that, in practice, a high degree of *hybridity* will probably arise (as companies make efforts to improve their environmental performance). In this regard, three final points can be made.

First, a sustainable food network accommodates the many different locally specific pathways that already exist, and that offer a portfolio of cases and examples for other areas – North and South, prosperous and low-income – from which to learn. This process will involve not a simple replication of "what worked elsewhere", however, but a need to ensure that new initiatives are designed for local circumstances, are socially inclusive, and engage people of all ages. Second, there is a key role to be played by local to mid-level government (town, borough, city and county councils and other authorities), where land-use planning generally resides. The opportunity to be associated with "sustainable" rather than "alternative" (and, by implication, oppositional) initiatives should be attractive for planners and politicians alike. Planning can then be used positively for the "greening" of urban space for food growing, local markets and other food-related public provision. Indeed, it has the power to halt, and even to reverse, the inexorable hollowing out of town and city centres as corporate retailers drive relocation to suburban shopping malls. Third, and finally, there can be no substitute for a greater level of social mobilisation around food. The involvement of local government must facilitate and support local civic action, not take it over, redirect it or seek to close it down. The development of food citizenship may be critical in helping to recover some public policy leverage over the mainstream food system. And for new food citizens, perhaps the best examples from which to learn are those popular movements for food sovereignty that are spreading in the South – movements that seek to reclaim their right to determine what is grown in order to meet the needs of local and regional markets. This need not be about creating autarchy, but is rather about re-establishing the rights of people to safe, healthy and ecologically sustainable production of food for all.

Glossary

Agri-commodities Agricultural products produced for the global market and whose value is established by the parameters of supply, demand and speculation.

Agri-food system Encompasses all those activities related to the production (cultivation, rearing, capture); processing (refining, manufacturing); distribution (transportation); sale/retail, preparation; and consumption of food. The food system embraces a more holistic and dynamic understanding, recognising complex relationships between different components, including feedback loops from consumption that can reshape aspects of production. In this it possesses a superior explanatory capability to the term *food chain*.

Agro-ecosystem Represents human manipulation and alteration of a natural ecosystem for the purpose of establishing agricultural production. Ecosystem properties, such as energy flows and nutrient cycling, are harnessed in such a way as to optimise productivity of the desired crop(s).

Alternative food networks Widely used term that embraces a variety of new arrangements linking together producers, consumers and other actors in more fluid and loosely bound sets of relations that together represent an alternative to more standardised food supply chains.

Appropriationism A process whereby activities and resources previously under the control of farmers and local artisans (manufacture of farm tools and mechanical aids, soil fertilisers and seeds) become the basis of an autonomous sector of innovation and control, leading to the accumulation of value by industrial interests at the expense of farmers.

Consumerism The fostering of desire to acquire material goods.

Distanciation A term possessing several different philosophical meanings, it is used here in a more prosaic sense to convey the way in which a perceptual separation is established between the realm of production and the site of consumption, such that it conceals the circumstances in which food may be produced.

Entitlements Can be regarded as rights of ownership or use to the products of land or labour, which determine people's access to food. For example, a farmer growing her own food is entitled to what she has grown (less any obligations to others). The entitlement of a labourer working for wages is constituted by what he can buy with those wages; in the absence of employment such entitlements collapse, leaving labouring families vulnerable to hunger.

Food chain A term commonly used by agri-food policy makers and business analysts to refer to the succession of stages from primary production to consumer purchases of food. Unlike a food systems approach, food chains tend toward a linear representation of discrete sequential stages that are frequently labelled as "farm to fork" or "field to plate".

Food security Conventionally, "food security exists when all people, at all times, have physical and economic access to sufficient, safe and nutritious food to meet their dietary needs and food preferences for an active and healthy life" (FAO 1997). Alternatively, community food security exists when "all residents obtain a safe, culturally acceptable, nutritionally adequate diet through a sustainable food system that maximizes community self-reliance, social justice, and democratic decision-making" (Winne n.d.).

Food sovereignty The right of peoples and sovereign states to democratically determine their own food and agricultural policies.

Global warming potential GWP is a measure of the ability of a greenhouse gas to trap heat in the atmosphere. This measure combines its efficiency to absorb and radiate heat, and its lifespan in the atmosphere. The GWP of products is calculated on the basis of the total greenhouse gas emissions arising from their production, use and disposal.

Horizontal consolidation A process whereby a company merges with or acquires another within the same sector, thus increasing their share of that particular market segment.

Productivist/productivism An approach to agricultural production that emphasises output or yield per unit area and effectively disregards all other considerations, such as the long-term consequences of resource depletion or social justice. Productivism represents the polar opposite to sustainability.

Relocalisation A vital aspect of alternative and sustainable food networks that seek to reconnect with the place of production, in contrast to the perceived "placelessness" of globalised food "from nowhere".

Substitutionism The capability developed by food manufacturers to interchange different raw materials as ingredients for their final food products, including the displacement of primary foods by industrial (synthetic) substitutes.

Sustainable consumption The use of goods and services for the satisfaction of basic needs and to enhance the quality of life, while minimising the use of natural resources and toxic materials as well as the emissions of waste and pollutants over the life cycle.

Sustainable food system Combining a concern with the present and future generations, an SFS is one that provides sufficient healthy food to meet current needs while ensuring the integrity of agro-ecosystems and the wider environment to meet the food needs of generations to come. It fosters greater local production and distribution in order to ensure that nutritious food is available, accessible and affordable to all. An SFS is built on humane practices, building equitable and just trading partnerships with distant producers, protecting farmers, workers and consumers.

Vertical integration A strategy of agri-food companies to develop commercial interests upstream and/or downstream of their existing core competence, usually by acquisition (take-over) of businesses in that area.

Bibliography

AEA (2007) 'Scoping studies to identify opportunities for improving resource use efficiency and for reducing waste through the food production chain', Report to Department for Environment, Food and Rural Affairs, London.

Allen, P. (ed.) (1993) *Food for the Future: Conditions and Contradictions of Sustainability*. New York: John Wiley & Sons.

Ambler-Edwards, S., Bailey, K., Kiff, A., Lang, T., Lee, R., Marsden, T., Simons, D., Tibbs, H. (2009) *Food Futures: Rethinking UK Strategy*, A Chatham House report. London: Royal Institute of International Affairs.

APA (2007) *Policy Guide on Community and Regional Food Planning*, Chicago, IL/Washington, DC: American Planning Association, www.planning.org/policyguides/adopted/food.htm

APHA (2007) 'Toward a healthy, sustainable food system', policy statement no. 200712. Washington, DC: American Public Health Association, www.apha.org/advocacy/policy/policysearch

Andersson, K., Ohlsson, T., Olsson, P. (1998) 'Screening life cycle assessment (LCA) of tomato ketchup: a case study', *Journal of Cleaner Production* 6: 277–88.

Atkins, P., Bowler, I. (2001) *Food in Society: Economy, Culture, Geography*. London: Arnold.

Bailey, I., Hopkins, R., Wilson, G. (2010) 'Some things old, some things new: the spatial representations and politics of change of the peak oil relocalisation movement', *Geoforum* 41, 4: 595–605.

Barrett, H., Browne, A., Ilberry, B. (2004) 'From farm to supermarket: the trade in fresh horticultural produce from sub-Saharan Africa to the United Kingdom', in A. Hughes and S. Reimer (eds) *Geographies of Commodity Chains*. London: Routledge, 19–38.

Battisti, D., Naylor, R. (2009) 'Historical warnings of future food insecurity with unprecedented seasonal heat', *Science* 323: 240–44.

BBC News (2004) Soft drink is purified tap water, Monday 1 March 2004. http://news.bbc.co.uk/2/hi/uk_news/3523303.stm

Bell, D., Valentine, G. (1997) *Consuming Geographies: We are Where we Eat*, London: Routledge.

Bellarby, J., Foereid, B., Hastings, A., Smith, P.(2008) *Cool Farming: Climate Impacts of Agriculture and Mitigation Potential*. Amsterdam: Greenpeace International.

Berlan, A., Dolan, C. (forthcoming) 'Rethinking Fairtrade's ethic of relationality: the case of cocoa producers', in M. Goodman and C. Sage (eds) *Food Transgressions: Making Sense of Contemporary Food Politics*. Aldershot, UK: Ashgate.

Biochar Research Centre (2009) 'What is biochar?', www.biochar.org.uk/what_is_biochar.php

Blair, D., Sobal, J. (2006) 'Luxus consumption: wasting food resources through overeating', *Agriculture and Human Values* 23: 63–74.

Blythman, J. (2004) *Shopped: The Shocking Power of British Supermarkets*. London: Fourth Estate.

Böge, S. (1995) 'The well-travelled yogurt pot: lessons for new freight transport policies and regional production', *World Transport Policy and Practice* 1, 1: 7–11.

Born, B., Purcell, M. (2006) 'Avoiding the local trap: scale and food systems in planning research', *Journal of Planning Education and Research* 26: 195–207.

Brander, K. (2007) 'Global fish production and climate change', *Proceedings of the National Academy of Sciences USA* 104, 50: 19709–14.

von Braun, J. (2005) 'The World Food Situation: An Overview', paper prepared for CGIAR AGM, Marrakech, December. Washington, DC: International Food Policy Research Institute.

—— (2007) 'The World Food Situation: New Driving Forces and Required Actions', working paper. Washington, DC: International Food Policy Research Institute.

von Braun, J., Diaz-Bonilla, E. (2008) 'Globalization of agriculture and food: causes, consequences, and policy implications', in J. von Braun and E. Diaz-Bonilla (eds) *Globalization of Food and Agriculture and the Poor*. New Delhi: Oxford University Press, 1–45.

Braxton Little, J. 2009 'The Ogallala Aquifer: saving a vital U.S. water source', *Scientific American* March, www.scientificamerican.com/article.cfm?id=the-ogallala-aquifer

Bridge, G. (2010) 'Geographies of peak oil: the other carbon problem', *Geoforum* 41: 523–30.

Bridger, R. (2008) 'Food air-freight, the global infrastructure expansion', unpublished paper. Brighton, UK: Food Ethics Council.

Brown, L. (2005) *Outgrowing the Earth: The Food Security Challenge in an Age of Falling Water Tables and Rising Temperatures*. London: Earthscan.

Bunyard, P. (2007) 'Climate and the Amazon', in H. Girardet (ed.) *Surviving the Century: Facing Climate Chaos and Other Global Challenges*. London: Earthscan, 77–101.

Busch, L., Bain, C. (2004) 'New! Improved? The transformation of the global agrifood system', *Rural Sociology* 69, 3: 321–46.

Buttimer, A., Stol, T. (2001) 'Flows of food and fuel in Germany, Ireland, the Netherlands and Sweden, 1960–90', in A. Buttimer (ed.) *Sustainable*

Landscapes and Lifeways: Scale and Appropriateness. Cork: Cork University Press, 79–103.

Cabinet Office (2008) *Food: An Analysis of the Issues*. London: Strategy Unit, Cabinet Office, UK government.

Campbell, C. (1997) *The Coming Oil Crisis*, Brentwood, Essex: Multi-Science Publishing and Petroconsultants S.A.

—— (2003) *The Essence of Oil and Gas Depletion*. Brentwood, Essex: Multi-Science Publishing.

Campbell, C., Laherrère, J. (1998) 'The end of cheap oil', *Scientific American* 278: 80–85.

Caraher, M., Dixon, P., Lang, T., Carr-Hill, R. (1998) 'Access to healthy foods: Part I. Barriers to accessing healthy foods: differentials by gender, social class, income and mode of transport', *Health Education Journal* 57: 191–201.

Carlsson-Kanyama, A., Lindén, A.-L. (2001) 'Trends in food production and consumption: Swedish experiences from environmental and cultural impacts', *International Journal of Sustainable Development* 4, 4: 392–406.

Carlsson-Kanyama, A., Ekström, M., Shanahan, H. (2003) 'Food and life cycle energy inputs: consequences of diet and ways to increase efficiency', *Ecological Economics* 44: 293–307.

Carson, R. (1962) *Silent Spring*. Boston, MA: Houghton Mifflin (UK edn 1963, London: Hamish Hamilton).

Chapagain, A., Hoekstra, A. (2008) 'The global component of freshwater demand and supply: an assessment of virtual water flows between nations as a result of trade in agricultural and industrial products', *Water International* 33, 1: 19–32.

CIAA (2007) *Managing Environmental Sustainability in the European Food and Drink Industries*. Brussels: Confederation of the Food and Drink Industries of the EU.

Clarke, W. (1996) 'The sustainability of sustenance: land and agricultural production in the Third World', in I. Douglas, R. Huggett, M. Robinson (eds) *Companion Encyclopaedia of Geography, The Environment and Humankind*. London: Routledge, 651–76.

Clay, E. (1997) *Food Security: A Status Review of the Literature*, Research Report. London: Overseas Development Institute.

Clay, E., Pillai, N., Benson, C. (1998) *The Future of Food Aid: A Policy Review*, Research Report. London: Overseas Development Institute

Clunies-Ross, T., Hildyard, N. (1992) *The Politics of Industrial Agriculture*. London: Earthscan.

Cole, M., Miele, M., Hines, P., Zokaei, K., Evans, B., Beale, J. (2009) 'Animal foods and climate change: shadowing eating practices', *International Journal of Consumer Studies* 33: 162–67.

Cook, I. (2004) 'Follow the thing: papaya', *Antipode* 36, 4: 642–64.

Cook, I., Crang, P.(1996) 'The world on a plate: culinary culture, displacement and geographical knowledges', *Journal of Material Culture* 1, 2: 131–53.

Corporate Accountability International (n.d.) '{Students} think outside the bottle', www.stopcorporateabuse.org/students-think-outside-bottle

Counihan, C., van Esterik, P. (eds) (2008) *Food and Culture: A Reader*, 2nd edn. Abingdon, UK: Routledge.

Crabb, P. 1996 'Water: confronting the critical dilemma', in I. Douglas, R. Huggett, M. Robinson (eds) *Companion Encyclopaedia of Geography, the Environment and Humankind*. London: Routledge, 526–52.

Davidson, A. (2006) *The Oxford Companion to Food* (2nd edn edited by T. Jaine). Oxford: Oxford University Press.

Defra (2005) *The Validity of Food Miles as an Indicator of Sustainable Development*. York: Food Statistics Branch, Department for Environment, Food and Rural Affairs.

—— (2007) *Food Transport Indicators to 2006 (revised): Experimental Statistics*. York: Food Statistics Branch, Department for Environment, Food and Rural Affairs.

—— (2008a) *Food Statistics Pocketbook 2008*. York: Food Statistics Branch, Department for Environment, Food and Rural Affairs.

—— (2008b) *PAS2050 Case Study: Applying PAS2050 to a Complex Product: Cottage Pie Ready Meal*. London: Department for Environment, Food and Rural Affairs.

—— (2008c) 'Ensuring the UK's food security in a changing world', Defra discussion paper. London: Food Statistics Branch, Department for Environment, Food and Rural Affairs.

—— (2009) *Food Statistics Pocketbook 2009*. York: Food Statistics Branch, Department for Environment, Food and Rural Affairs.

DeRose, L., Millman, S. (1998) 'Introduction', in L. DeRose, E. Messer, S. Millman (eds) *Who's Hungry? And How Do We Know? Food Shortage, Poverty and Deprivation*. Tokyo: United Nations University Press.

Devereux, S., Edwards, J. (2004) 'Climate change and food security', *IDS Bulletin* 35, 3: 22–30.

Diamond, J. (1998) *Guns, Germs and Steel: A Short History of Everybody for the Last 13,000 years*. London: Vintage.

Dinham, B. (2005) 'Corporations and pesticides', in J. Pretty (ed.) *The Pesticide Detox: Towards a More Sustainable Agriculture*. London: Earthscan, 55–69.

Dorward, A., Poulton, C. (2008) 'The Global Fertiliser Crisis and Africa', working paper. Sussex: Future Agricultures Consortium.

Drèze, J., Sen, A. (1989) *Hunger and Public Action*. Oxford: Clarendon Press.

DuPuis, M., Goodman, D. (2005) 'Should we go "home" to eat? Towards a reflexive politics of localism', *Journal of Rural Studies* 21: 359–71.

Dyson, T. (1996) *Population and Food: Global Trends and Future Prospects*. London: Routledge.

Earth Policy Institute (2007) 'Bottled water boycotts: back-to-the-tap movement gains momentum', *Plan B Update*, www.earth-policy.org/plan_b_updates/2007/update68.

EC SCAR (2009) *Second SCAR Foresight Exercise: New Challenges for Agricultural Research*. Luxembourg: European Commission, Standing Committee on Agricultural Research.

Eco-Logica Ltd (2003) *Wisemoves: The Potential to Reduce Greenhouse Gas Emissions through Localisation and Co-operation*, Final Report. Lancaster, UK: Eco-Logical Ltd.

Edwards-Jones, G., Plassmann, K., York, E., Hounsome, B., Jones, D., Milà, I., Canals, L. (2009) 'Vulnerability of exporting nations to the development of a carbon label in the United Kingdom', *Environmental Science and Policy* 12: 479–90.

Environmental Integrity Project (2008) T*ar Sands: Feeding U.S. Refinery Expansions with Dirty Fuel*. Washington, DC: Environmental Integrity Project, http://environmentaldefence.ca/sites/default/files/report_files/EIPTar%20Sand%20Report_FINAL.pdf

Eurostat (2008) *EU Economic Data Pocketbook, 2/2008*. Luxembourg: European Commission.

Evans, A. (2009) 'The feeding of the nine billion: global food security for the 21st century', Chatham House report. London: Royal Institute of International Affairs.

Fairtrade Foundation (2009) Sales of Fairtrade certified products in the UK. www.fairtrade.org.uk/what_is_fairtrade/facts_and_figures.aspx

Feagan, R. (2007) 'The place of food: mapping out the "local" in local food systems', *Progress in Human Geography* 31, 1: 23–42.

Feehan, J. (2003) *Farming in Ireland: History, Heritage and Environment*. Dublin: University College Dublin, Faculty of Agriculture.

Fernandes, E., Pell, A., Uphoff, N. (2005) 'Rethinking agriculture for new opportunities', in J. Pretty (ed.) *The Earthscan Reader in Sustainable Agriculture*. London: Earthscan, 321–40.

Fernandez-Armesto, F. (2001) *Food: A History*. London: Macmillan.

Fiala, N. (2008) 'Meeting the demand: an estimation of potential future greenhouse gas emissions from meat production', *Ecological Economics* 67: 412–19.

Fishman, C. (2006) *The Wal-Mart Effect: How an Out-of-Town Superstore Became a Superpower*. London: Allen Lane.

FISS (2006) 'Food Industry Sustainability Strategy', London: Food Industry Sustainability Strategy, Department for Environment, Food and Rural Affairs.

Folke, C. (2006) 'Resilience: the emergence of a perspective for social–ecological systems analyses', *Global Environmental Change* 16: 253–67.

FAO (1997) *Report of the World Food Summit, 13–17 November 1996, Part One*. Rome: Food and Agriculture Organization of the United Nations.

—— (2006) *World Agriculture: Towards 2030/2050. Interim Report: Prospects for Food, Nutrition, Agriculture and Major Commodity Groups*. Rome:

Global Perspective Studies Unit, Food and Agriculture Organization of the United Nations.

—— (2008) *The State of Food Insecurity in the World 2008: High Food Prices and Food Security – Threats and Opportunities*. Rome: Food and Agriculture Organization of the United Nations.

—— (n.d.) FAOSTAT, http://faostat.fao.org

Food & Water Watch (2007) *Take Back the Tap: Why Choosing Tap Water over Bottled Water is Better for Your Health, Your Pocketbook, and the Environment*. Washington, DC: Food & Water Watch, www.foodandwaterwatch.org/water/bottled/pubs/reports/take-back-the-tap

Forum for a New World Governance (2010) 'People's Food Sovereignty Statement', www.world-governance.org/spip.php?article70

Foster, C. (2008) 'Food Product LCAs: A Review for Policymakers. Presentation to 2008 Agricultural Economics Society Conference, https://statistics.defra.gov.uk/esg/conference/aes2008/default.asp [URL no longer available]

—— (2011) personal communication.

Fresco, L. (2009) 'Challenges for food system adaptation today and tomorrow', *Environmental Science and Policy* 12, 4: 378–85.

Friedberg, S. (2004) 'The ethical complex of corporate food power', *Environment and Planning D: Society and Space* 22: 513–31.

Friedmann, H. (2009) 'Discussion: moving food regimes forward: reflections on symposium essays', *Agriculture and Human Values* 26: 335–44.

Galeano, E. (1973) *Open Veins of Latin America: Five Centuries of the Pillage of a Continent*. New York: Monthly Review Press.

Gallagher, K., Ooi, P., Mew, T., Borromeo, E., Kenmore, P., Ketelaar, J.-W. (2005) 'Ecological basis for low-toxicity integrated past management (IPM) in rice', in J. Pretty (ed.) *The Earthscan Reader in Sustainable Agriculture*. London: Earthscan, 206–20.

Garnett, T. (2007) 'Food refrigeration: what is the contribution to greenhouse gas emissions and how might emissions be reduced?', Food Climate Research Network working paper. Guildford, UK: University of Surrey.

—— (2008) 'Cooking up a storm: food, greenhouse gas emissions and our changing climate', Food Climate Research Network working paper. Guildford, UK: University of Surrey.

—— (2009) 'Livestock-related greenhouse gas emissions: impacts and options for policy makers', *Environmental Science and Policy* 12: 491–503.

Gehlhar, M., Regmi, A. (2005) 'Factors shaping global food markets', in A. Regmi and M. Gehlhar (eds) *New Directions in Global Food Markets*, Agriculture Information Bulletin No. 794. Washington, DC: United States Department of Agriculture, 5–17.

Godfray, H.C., Crute, I., Haddad, L., Lawrence, D., Muir, J., Nisbett, N., Pretty, J., Robinson, S., Toulmin, C., Whiteley, R. (2010) 'The future of the global food system', *Philosophical Transactions of the Royal Society B* 365: 2769–77.

Goodman, D. (2003) 'Editorial: the quality "turn" and alternative food practices: reflections and agenda', *Journal of Rural Studies* 19, 1: 1–7.

Goodman, D., Redclift, M. (1991) *Refashioning Nature: Food, Ecology and Culture*. London: Routledge.

Goodman, M., Maye, D., Holloway, L. (2010a) 'Guest editorial: ethical foodscapes? Premises, promises and possibilities', *Environment and Planning A* 42: 1782–96.

Goodman, M., Goodman, D., Redclift, M. (eds) (2010b) *Consuming Space: Placing Consumption in Perspective*. Aldershot, UK: Ashgate.

Goody, J. (1997) 'Industrial food: towards the development of a world cuisine', in C. Counihan and P. van Esterik (eds) *Food and Culture: A Reader*. New York: Routledge, 338–56.

Government of Alberta (2010) 'Oil Sands', www.energy.gov.ab.ca/OilSands/oilsands.asp

Green, K., Foster, C. (2005) 'Give peas a chance: transformations in food consumption and production systems', *Technological Forecasting and Social Change* 72: 663–79.

Gregory, P., Ingram, J., Brklacich, M. (2005) 'Climate change and food security', *Philosophical Transactions of the Royal Society B* 360: 2139–2148.

Gruppi di Acquisto Solidale (2010) 'What's a G.A.S?', www.retegas.org/index.php

Gussow, J.D. (1994) 'Ecology and vegetarian considerations: does environmental responsibility demand the elimination of livestock?', *American Journal of Clinical Nutrition* 59 (suppl.): 1110–16.

Harkin, T. (2004) 'Economic Concentration and Structural Change in the Food and Agriculture Sector: Trends, Consequences and Policy Options', Paper prepared for the Committee on Agriculture, Nutrition, and Forestry, United States Senate, Washington, DC.

Harvey, G. (2006) *We Want Real Food: Why our Food is Deficient in Minerals and Nutrients – and What we Can do About It*. London: Constable.

Harvey, M., McMeeking, A, Warde, A. (eds) (2004) *Qualities of Food*. Manchester: Manchester University Press.

van Hauwermeiren, A., Coene, H., Engelen, G., Mathijs, E. (2007) 'Energy lifecycle inputs in food systems: a comparison of local versus mainstream cases', *Journal of Environmental Policy and Planning* 9, 1: 31–51.

Hawken, P., Lovins, A., Lovins, L. (1999) *Natural Capitalism: The Next Industrial Revolution*. London: Earthscan.

Hawkes, C. (2006) 'Uneven dietary development: linking the policies and processes of globalisation with the nutrition transition, obesity and diet-related chronic diseases', *Globalization and Health* 2: 4.

——— (2007) 'Agro-food industry growth and obesity in China: what role for regulating food advertising and promotion and nutrition labelling?', *Obesity Reviews* 9 (suppl. 1): 151–61.

——— (2008) 'Dietary implications of supermarket development: a global perspective', *Development Policy Review* 26, 6: 657–92.

Heasman, M., Mellentin, J. (2001) *The Functional Foods Revolution: Healthy People, Healthy Profits?* London: Earthscan.

Heffernan, W. (1999) 'Consolidation in the food and agriculture system', report to the National Farmers Union. Columbia, MO: Department of Rural Sociology, University of Missouri.

Hendrickson, M., Heffernan, W. (2007) 'Concentration of agricultural markets', unpublished paper. Columbia, MO: Department of Rural Sociology, University of Missouri.

Hinton, E., Goodman, M. (2010) 'Sustainable consumption: developments, considerations and new directions', in M. Redclift, G. Woodgate (eds) *The International Handbook of Environmental Sociology*, 2nd edn. Cheltenham, UK: Edward Elgar, 245–61.

Hoddinott, J. (1999) *Operationalizing Household Food Security in Development Projects: An Introduction*, Technical Guide No.1. Washington, DC: International Food Policy Research Institute.

Hoekstra, A., Chapagain, A. (2007) 'Water footprints of nations: water use by people as a function of their consumption pattern', *Water Resource Management* 21: 35–48.

——— (2008) *Globalisation of Water: Sharing the Planet's Freshwater Resources*. Oxford: Blackwell.

Holt-Giménez, E., Patel, R. (2009) *Food Rebellions! Crisis and the Hunger for Justice*. Cape Town: Pambazuka Press.

Holt-Giménez, E., Bailey, I., Sampson, D. (2007) 'Fair to the last drop: the corporate challenges to fair trade coffee', Development Report 17. Oakland, CA: Institute for Food and Development Policy.

Höök, M., Hirsch, R., Aleklett, K. (2009) 'Giant oil field decline rates and their influence on world oil production', *Energy Policy* 37: 2262–72.

Howard, A. (1943) *An Agricultural Testament*. New York and London: Oxford University Press (first published in England, 1940). http://journeytoforever. org/farm_library/howardAT/ATtoc.html

IAASTD (2009) *Synthesis Report: A Synthesis of the Global and Sub-Global IAASTD Reports*. Washington, DC: Island Press/International Assessment of Agricultural Knowledge, Science and Technology for Development.

IGD (2007) 'Global Retailing 2007', Watford, UK: IGD. www.igd.com

International Rivers (2009) 'The Three Gorges Dam'. http://internationalrivers. org/china/three-gorges-dam

IPCC (2007) *Climate Change 2007: Synthesis Report*. Contribution of Working Groups I, II, and III to the Fourth Assessment Report of the Intergovernmental Panel on Climate Change, edited by Core Writing

Team, R.K. Pachauri, A. Reisinger. Geneva: Intergovernmental Panel on Climate Change.

Jackson, P., Russell, P., Ward, N. (2007) 'The appropriation of "alternative", discourses by "mainstream" food retailers', in D. Maye, L. Holloway, M. Kneafsey (eds) *Alternative Food Geographies: Representation and Practice.* Oxford: Elsevier, 309–30.

Jacobson, M. (2006) *Six Arguments for a Greener Diet: How a More Plant-Based Diet Could Save Your Health and the Environment.* Washington, DC: Centre for Science in the Public Interest.

Jarosz, L. (2008) 'The city in the country: growing alternative food networks in metropolitan areas', *Journal of Rural Studies* 24: 231–44.

Jones, A. (2001) *Eating Oil: Food Supply in a Changing Climate.* London: Sustain and Elm Farm Research Centre.

Keyzer, M., Merbis, M., Pavel, I., van Wesenbeeck, C. (2005) 'Diet shifts towards meat and the effects on cereal use: can we feed the animals in 2030?', *Ecological Economics* 55: 187–202.

Khan, S., Hanjra, M. (2008) 'Footprints of water and energy inputs in food production – global perspectives', *Food Policy* 34: 130–40.

Kickbusch, I. (2010) *The Food System: A Prism of Present and Future Challenges for Health Promotion and Sustainable Development.* Bern/Lausanne: Health Promotion Switzerland. www.iuhpeconference.net/ downloads/en/Programme/White-Paper--The-Food-System.pdf

Kimbrell, A. (ed.) (2002) *Fatal Harvest: The Tragedy of Industrial Agriculture.* Sausalito, CA: Foundation for Deep Ecology/Island Press.

Kirwan, J., Foster, C. (2007) 'Public sector food procurement in the United Kingdom: examining the creation of an "alternative" and localised network in Cornwall', in D. Maye, L. Holloway, M. Kneafsey (eds) *Alternative Food Geographies: Representation and Practice.* Oxford: Elsevier, 185–201.

Kishi, M. (2005) 'The health impacts of pesticides: what do we now know?', in J. Pretty (ed.) *The Pesticide Detox: Towards a More Sustainable Agriculture.* London: Earthscan, 23–38.

Kneafsey, M. (2010) 'The region in food – important or irrelevant?', *Cambridge Journal of Regions, Economy and Society* 3: 177–90.

Kurlansky, M. (1999) *Cod: A Biography of the Fish that Changed the World.* London: Vintage Books.

—— (2003) *Salt: A World History.* London: Vintage Books.

Lancet (2009) 'Managing the health effects of climate change', *Lancet* 373: 1693–1733.

Lang, T. (2010) 'From "value-for-money" to "values-for-money"? Ethical food and policy in Europe', *Environment and Planning A* 42: 1814–32.

Lang, T., Caraher, M. (1998) 'Access to healthy foods: Part II. Food poverty and shopping deserts: what are the implications for health promotion policy and practice?', *Health Education Journal* 57: 202–11.

Lang, T., Heasman, M. (2004) *Food Wars: The Global Battle for Mouths, Minds and Markets*. London: Earthscan.

Lang, T., Barling, D., Caraher, M. (2009) *Food Policy: Integrating Health, Environment and Society*. Oxford: Oxford University Press.

Lappé, F.M., Collins, J. (1982) *Food First*. London: Abacus.

Lawrence, F. (2004) *Not on the Label: What Really Goes into the Food on Your Plate*. London: Penguin.

—— (2008) *Eat Your Heart Out: Why the Food Business is Bad for the Planet and your Health*. London: Penguin.

Lawrence, G., Lyons, K., Wallington, T. (eds) 2010 *Food Security, Nutrition and Sustainability*. London: Earthscan.

Leach, G. (1976) *Energy and Food Production*. Guildford, UK: IPC Press.

Leach, M., Scoones, I., Stirling, A. (2010) *Dynamic Sustainabilities: Technology, Environment, Social Justice*. London: Earthscan.

Leggett, J. (2005) *Half Gone: Oil, Gas, Hot Air and the Global Energy Crisis*. London: Portobello Books.

Lucas, C., Jones, A., Hines, C. (2006) 'Fuelling a Food Crisis? The Impact of Peak Oil on Food Security', report for The Greens/European Free Alliance in the European Parliament.

Lyson, T. (2004) *Civic Agriculture: Reconnecting Farm, Food and Community*. University Park, PA: Penn State University Press.

Magdoff, F., Foster, J.B., Buttel, F. (2000) 'An overview', in F. Magdoff, J.B. Foster, F. Buttel (eds) *Hungry for Profit: The Agribusiness Threat to Farmers, Food and the Environment*. New York: Monthly Review Press, 7–21.

Mannion, A. (1995) *Agriculture and Environmental Change: Temporal and Spatial Dimensions*. Chichester, UK: John Wiley.

Marsden, T. (2003) *The Condition of Rural Sustainability*. Assen, the Netherlands: Royal van Gorcum.

Maxey, L. (2007) 'From "alternative" to "sustainable" food', in D. Maye, L. Holloway, M. Kneafsey (eds) *Alternative Food Geographies: Representation and Practice*. Oxford: Elsevier, 55–75.

Maye, D., Holloway, L., Kneafsey, M. (eds) (2007) *Alternative Food Geographies: Representation and Practice*. Oxford: Elsevier.

Mazoyer, M., Roudart, L.(2006) *A History of World Agriculture: From the Neolithic to the Current Crisis*. London: Earthscan.

McClintock, N. (2010) 'Why farm the city? Theorizing urban agriculture through a lens of metabolic rift', *Cambridge Journal of Regions, Economy and Society* 3: 191–297.

McGuire, V. (2007) 'Changes in water levels and storage in the High Plains aquifer, predevelopment to 2005', United States Geological Service report, http://pubs.usgs.gov/fs/2007/3029

McMichael, A. (2005) 'Integrating nutrition with ecology: balancing the health of humans and biosphere', *Public Health Nutrition* 8, 6A: 706–15.

McMichael, P. (2000) 'The power of food', *Agriculture and Human Values* 17: 21–33.

—— (2005) 'Global development and the corporate food regime', in F. Buttel and P. McMichael (eds) *New Directions in the Sociology of Global Development*, Research in Rural Sociology and Development Vol. 11. Amsterdam: Elsevier, 265–99.

Milestad, R., Westberg, L., Geber, U., Bjorklund, J. (2010) 'Enhancing adaptive capacity in food systems: learning at farmers' markets in Sweden', *Ecology and Society* 15, 3.

Millennium Ecosystem Assessment (2005a) *Ecosystems and Human Well-being: Synthesis*. Washington, DC: Island Press. www.maweb.org/en/Reports.aspx

—— (2005b) *Ecosystems and Human Well-being: Current State and Trends*. Washington, DC: Island Press. www.maweb.org/en/Reports.aspx

—— (2005c) *Ecosystems and Human Well-being. Policy Responses*. Washington, DC: Island Press. www.maweb.org/en/Reports.aspx

Millstone, E., Lang, T. (2008) *The Atlas of Food: Who Eats What, Where and Why*, 2nd edn. London: Earthscan.

Mintz, S. (2008) 'Time, sugar, and sweetness', in C. Counihan, P. van Esterik (eds) *Food and Culture: A Reader*, 2nd edn. Abingdon, UK: Routledge, 91–103.

Molle, F., Mollinga, P., Meinzen-Dick, R. (2008) 'Water, politics and development: Introducing water alternatives', *Water Alternatives* 1, 1: 1–6.

Morgan, K. (2008) 'Greening the realm: sustainable food chains and the public plate', *Regional Studies* 42, 9: 1237–50.

—— (2009) 'Feeding the city: the challenge of urban food planning', *International Planning Studies* 14, 4: 341–48.

—— (2010) 'Local and green, global and fair: the ethical foodscape and the politics of care', *Environment and Planning A* 42: 1852–67.

Morgan, K., Sonnino, R. (2010) 'The urban foodscape: world cities and the new food equation', *Cambridge Journal of Regions, Economy and Society* 3: 209–24.

Morgan, K., Marsden, T., Murdoch, J. (2006) *Worlds of Food: Place, Power and Provenance in the Food Chain*. Oxford: Oxford University Press.

Mouawad, J. (2008) 'Oil price rise fails to open tap', *The New York Times*, 29 April 2008, www.nytimes.com/2008/04/29/business/worldbusiness/29oil.html

Mougeot, L. (2005) 'Introduction', in L. Mougeot (ed.) *Agropolis: The Social, Political and Environmental Dimensions of Urban Agriculture*. London and Ottawa: Earthscan and IDRC, 1–29.

NASA (2009) 'Mississippi Dead Zone', www.nasa.gov/vision/earth/environment/dead_zone.html

Naylor, R., Goldburg, R., Primavera, J., Kautsky, N., Beveridge, M., Clay, J., Folke, C., Lubchenco, J., Mooney, H., Troell, M. (2000) 'Effect of aquaculture on world fish supplies', *Nature* 405: 1017–24.

Nestle, M. (2002) *Food Politics: How the Food Industry Influences Nutrition and Health.* Berkeley, CA: University of California Press.

—— (2003) *Safe Food: Bacteria, Biotechnology and Bioterrorism.* Berkeley, CA: University of California Press.

New Agriculturist (2009) 'Kenyan horticulture: weathering the political storm?', www.new-ag.info/08/02/develop/dev1.php

Oram, J. (2007) 'Efficiency or exploitation? Supermarket scale economies blur into abuse of market power', *Food Ethics* 2, 2: 6–7.

Page, B. (1997) 'Restructuring pork production, remaking rural Iowa', in D. Goodman, M. Watts (eds) *Globalising Food: Agrarian Questions and Global Restructuring.* London: Routledge, 133–57.

Page, P., White, R. (2000) 'Disposal of used packaging', in J. Dalzell (ed.) *Food Industry and the Environment in the European Union: Practical Issues and Cost Implications.* Gaithersburg, MD: Aspen, 291–334.

Parfitt, J., Barthel, M., Macnaughton, S. (2010) 'Food waste within food supply chains: quantification and potential for change to 2050', *Philosophical Transactions of the Royal Society B* 365: 3065–81.

Parrott, N., Wilson,N., Murdoch, J. (2002) 'Spatializing quality: regional protection and the alternative geography of food', *European Urban and Regional Studies* 9, 3: 241–61.

Patel, R. (2007) *Stuffed and Starved: Markets, Power and the Hidden Battle for the World Food System.* London: Portobello.

Pauly, D., Christensen, V., Guenette, S., Pitcher, T., Sumaila, R., Walters, C., Watson, R., Zeller, D. (2002) 'Toward sustainability in world fisheries', *Nature* 418: 689–95.

Peattie, K and Collins, A. (2009) 'Guest editorial: Perspectives on sustainable consumption', *International Journal of Consumer Studies* 33: 107–12.

Pew Centre (2009) *Key Scientific Developments Since the IPCC Fourth Assessment Report*, Science Brief 2. Arlington, VA: Pew Centre on Global Climate Change, www.pewclimate.org/brief/science-developments/June2009

Pfeiffer, D.A. (2006) *Eating Fossil Fuels: Oil, food and the Coming Crisis in Agriculture.* Gabriola Island, BC, Canada: New Society Publishers.

Pimentel, D. (2006) 'Soil erosion: a food and environmental threat', *Environment, Development and Sustainability* 8: 119–37.

Pimentel, D., Pimentel, M. (2003) 'Sustainability of meat-based and plant-based diets and the environment', *American Journal of Clinical Nutrition* 78 (suppl.): 660–63.

—— (2008) *Food, Energy and Society*, 3rd edn. Boca Raton, FL: CRC Press.

Pimentel, D., Shanks, R., Rylander, J. (2008) 'Energy use in fish and aquacultural production', in D. Pimentel, M. Pimentel (eds) *Food, Energy and Society.* Boca Raton, FL: CRC Press, 77–97.

Pinkerton, T., Hopkins, R. (2009) *Local Food: How to Make it Happen in your Community.* Totnes, UK: Green Books.

van der Ploeg, J.D. (2008) *The New Peasantries: Struggles for Autonomy and Sustainability in an Era of Empire and Globalisation*. London: Earthscan.

Polaris Institute (2010) 'Tar Sands Watch', www.tarsandswatch.org

Policy Commission on the Future of Farming and Food (2002) *Farming and Food: A Sustainable Future*. London: Cabinet Office.

Pollan, M. (2006) *The Omnivore's Dilemma: The search for a Perfect Meal in a Fast-food World*. London: Bloomsbury.

Poortinga, W. (2006) 'Perceptions of the environment, physical activity, and obesity', *Social Science and Medicine* 63: 2835–46.

Popkin, B. (2005) 'Using research on the obesity pandemic as a guide to a unified vision of nutrition', *Public Health Nutrition* 8(6A): 724–29.

—— (2006) 'Global nutrition dynamics: the world is shifting rapidly toward a diet linked with non-communicable diseases', *American Journal of Clinical Nutrition* 84, 2: 289–98.

Population Reference Bureau (2008) *2008 World Population Data Sheet*. Washington, DC: Population Reference Bureau, www.prb.org/pdf08/08WPDS_Eng.pdf

Pothukuchi, K. (2009) 'Community and regional food planning: building institutional support in the United States', *International Planning Studies* 14, 4: 349–67.

Pretty, J. (1995) *Regenerating Agriculture: Policies and Practice for Sustainability and Self-Reliance*, London: Earthscan.

—— (2002) *Agri-culture: Reconnecting People, Land and Nature*. London: Earthscan.

—— (ed.) (2005a) *The Earthscan Reader in Sustainable Agriculture*. London: Earthscan.

—— (2005b) *The Pesticide Detox: Towards a More Sustainable Agriculture*. London: Earthscan

—— (2007) *The Earth Only Endures: On Reconnecting with Nature and our Place in it*. London: Earthscan.

Pretty, J., Hine, R. (2005) 'Pesticide use and the environment', in J. Pretty (ed.) *The Pesticide Detox: Towards a More Sustainable Agriculture*. London: Earthscan, 1–22.

Pretty, J., Ball, A., Lang, T., Morison, J. (2005) 'Farm costs and food miles: an assessment of the full cost of the UK weekly food basket', *Food Policy* 30: 1–19.

Pretty, J., Sutherland, W., Ashby, J., Auburn, J., Baulcombe, D., Bell, M., Bentley, J., Bickersteth, S., Brown, K., Burke, J., Campbell, H., Chen, K., Crowley, E., Crute, I., Dobbelaere, D., Edwards-Jones, G., Funes-Monzote, F., Godfray, C., Griffon, M., Gypmantisiri, P., Haddad, L., Halavatau, S., Herren, H., Holderness, M., Izac, A-M., Jones, M., Koohafkan, P., Lal, R., Lang, T., McNeely, J., Mueller, A., Nisbett, N., Noble, A., Pingali, P., Pinto, Y., Rabbinge, R., Ravindranath, N.H., Rola, A., Roling, N., Sage, C., Settle, W., Sha, J.M., Shiming, L., Simons, T., Smith, P., Strzepeck, K.,

Swaine, H., Terry, E., Tomich, T., Toulmin, C., Trigo, E., Twomlow, S., Vis, J.K., Wilson, J., Pilgrim, S. (2010) 'The top 100 questions of importance to the future of global agriculture', *International Journal of Agricultural Sustainability* 8, 4: 219–36.

Reardon, T., Berdegué, J., Timmer, P. (2005) 'Supermarketization of the "emerging markets" of the Pacific Rim: development and trade implications', *Journal of Food Distribution Research* 36, 1: 3–12.

Redclift, M. (1994) 'Sustainable development: economics and the environment', in M. Redclift, C. Sage (eds) *Strategies for Sustainable Development: Local Agendas for the Southern Hemisphere*. Chichester, UK: John Wiley, 17–34.

Regmi, A., Gelhar, M. (eds) (2005) *New Directions in Global Food Markets*, Agriculture Information Bulletin 794. Washington, DC: US Department of Agriculture, Economic Research Service.

Reynolds, B. (2009) 'Feeding a world city: the London food strategy', *International Planning Studies* 14, 4: 417–24.

Riches, G. (1997) 'Hunger and the welfare state: comparative perspectives', in G. Riches (ed.) *First World Hunger: Food Security and Welfare Politics*. Basingstoke, UK: Macmillan, 1–13.

Rifkin, J. (1992) *Beyond Beef: The Rise and Fall of the Cattle Culture*. London: Thorsons.

Roberts, P. (2008) *The End of Food: The Coming Crisis in the World Food Industry*. London: Bloomsbury.

Rodell, M., Velicogna, I., Famiglietti, J. (2009) 'Satellite-based estimates of groundwater depletion in India', *Nature* 460: 999–1002.

Rodwan, J. 2009 'Confronting challenges: U.S. and International Bottled Water Developments and Statistics for 2008', *Bottled Water Reporter* April/May 2009, www.bottledwater.org

Röling, N. (2005) 'The human and social dimensions of pest management for agricultural sustainability', in J. Pretty (ed.) *The Pesticide Detox: Towards a More Sustainable Agriculture*. London: Earthscan.

Roth, D., Warner, J. (2008) 'Virtual water: virtuous impact? The unsteady state of virtual water', *Agriculture and Human Values* 25: 257–70.

Royal Society (2009) *Reaping the Benefits: Science and the Sustainable Intensification of Global Agriculture*, Policy document 11/09. London: The Royal Society.

Sage, C. (2003) 'Social embeddedness and relations of regard: alternative "good food" networks in South-West Ireland', *Journal of Rural Studies* 19, 1: 47–60.

—— (2007) 'Trust in markets: economies of regard and spaces of contestation in alternative food networks', in J. Cross, A. Morales (eds) *Street Trade: Commerce in a Globalising World*. London: Routledge, 147–63.

—— (2010) 'Re-imagining the Irish foodscape', *Irish Geography* 43, 2: 93–104.

Satterthwaite, D., McGranahan, G., Tacoli, C. (2010) 'Urbanization and its implications for food and farming', *Philosophical Transactions of the Royal Society B* 365: 2809–20.

Schlosser, E. (2001) *Fast Food Nation: What the All-American Meal is Doing to the World*. London: Penguin.

Scrinis, G., Lyons, K. (2007) 'The emerging nano-corporate paradigm: nanotechnology and the transformation of nature, food and agri-food systems', *International Journal of Sociology of Food and Agriculture* 15, 2: 22–44.

Selke, S. (2000) 'Packaging options', in J. Dalzell (ed.) *Food Industry and the Environment in the European Union: Practical Issues and Cost Implications*. Gaithersburg, MD: Aspen, 253–90.

Sen, A. (1981) *Poverty and Famines: An Essay on Entitlement and Deprivation*. Oxford: Clarendon Press.

Shurtleff, D.S., Burnett, H.S. (2008) 'Brazil's energy plan examined', *The Washington Times* 7 May 2008, www.washingtontimes.com/news/2008/may/7/brazils-energy-plan-examined

Sibbel, A. (2007) 'The sustainability of functional foods', *Social Science and Medicine* 64: 554–61.

Singer, P., Mason, J. (2006) *Eating: What we Eat and why it Matters*. London: Arrow Books.

Smil, V. (2000) *Feeding the World: A Challenge for the Twenty-first Century*. Cambridge, MA: MIT Press.

Smith, P., D. Martino, Z. Cai, D. Gwary, H. Janzen, P. Kumar, B. McCarl, S. Ogle, F. O',Mara, C. Rice, B. Scholes, O. Sirotenko (2007) 'Agriculture', in B. Metz, O.R. Davidson, P.R. Bosch, R. Dave, L.A. Meyer (eds) *Climate Change 2007: Mitigation*, Contribution of Working Group III to the Fourth Assessment Report of the Intergovernmental Panel on Climate Change. Cambridge: Cambridge University Press.

Sonnino, R. (2009) 'Feeding the city: towards a new research and planning agenda', *International Planning Studies* 14, 4: 425–35.

—— (2010) 'Escaping the local trap: insights on relocalisation from school food reform', *Journal of Environmental Policy and Planning* 12, 1: 23–40.

Steedman, P., Falk, T. (2009) 'From A to B: a snapshot of the UK food distribution system', Brighton, UK: Food Ethics Council.

Steinfeld, H., Gerber, P., Wassenaar, T., Castel, V., Rosales, M., de Haan, C. (2006) *Livestock's Long Shadow*. Rome: Food and Agriculture Organization, www.fao.org/docrep/010/a0701e/a0701e00.HTM

Stokstad, E. (2010) 'Could less meat mean more food?', *Science*, 12 February, 810–11.

Stuart, T. (2009) *Waste: Uncovering the Global Food Scandal*. London: Penguin.

Susskind, L. (1994) *Environmental Diplomacy: Negotiating More Effective Global Agreements*. Oxford: Oxford University Press.

Tannahill, R. (1988) *Food in History*. London: Penguin.

Tansey, G., Worsley, T. (1995) *The Food System: A Guide*. London: Earthscan.

Thompson, J., Scoones, I. (2009) 'Addressing the dynamics of food systems: an emerging agenda for social science research', *Environmental Science and Policy* 12, 4: 386–97.

Thompson, J., Millstone, E., Scoones, I., Ely, A., Marshall, F., Shah, E., Stagl, S. (2007) *Agri-food System Dynamics: Pathways to Sustainability in an Era of Uncertainty*, STEPS Working Paper 4. Brighton: STEPS Centre, Institute of Development Studies, University of Sussex.

Thompson, P. (2010) 'What sustainability is (and what it isn't)', in S. Moore (ed.) *Pragmatic Sustainability: Theoretical and Practical Tools*. Abingdon, UK: Routledge, pp. 16–29.

Tregear, A. (2003) 'From Stilton to Vimto: using food history to rethink typical products in rural development', *Sociologia Ruralis* 43: 91–107.

Tudge, C. (2003) *So Shall We Reap: What's Gone Wrong with the World's Food and How to Fix it*. London: Penguin.

Tukker, A., Huppes, G., Guinée, J., Heijungs, R., de Koning, A., van Oers, L., Suh, S., Geerken, T., Van Holderbeke, M., Jansen, B., Nielsen, P. (2006) *Environmental Impact of Products (EIPRO): Analysis of the Life Cycle Environmental Impacts Related to the Total Final Consumption of the EU25*. Brussels: Institute for Prospective Technological Studies/European Science and Technology Observatory, European Commission Joint Research Centre, http://ec.europa.eu/environment/ipp/pdf/eipro_report.pdf

UNEP (2008) *Kick the Habit: A UN Guide to Climate Neutrality*, Nairobi: United Nations Environment Programme, www.unep.org/publications/ebooks/kick-the-habit/Pdfs.aspx

UNEP-GRID (2010) 'The Aral Sea'. Nairobi: United Nations Environment Programme, http://enrin.grida.no/htmls/aralsoe/aralsea/english/arsea/arsea.htm

Vidal, J. (1997) *McLibel: Burger Culture on Trial*. London: Macmillan.

Vorley, B. (2003) *Food, Inc. Corporate Concentration from Farm to Consumer*. London: UK Food Group/International Institute for Environment and Development.

—— (2008) 'Air freight and Africa: Trading off environment and development?', presentation to Food Ethics Council workshop on Air-Freighted Food, 4 April 2008, Brighton, UK.

Wackernagel, M., Rees, W. (1996) *Our Ecological Footprint: Reducing Human Impact on the Earth*, Vancouver, BC, Canada: New Society Publishers.

Walt, V. (2008) 'The world's growing food-price crisis', *Time*, 27 February 2008, www.time.com/time/world/article/0,8599,1717572,00.html

Warshall, P. (2002) 'Tilth and technology: the industrial redesign of our nation's soils', in A. Kimbrell (ed.) *Fatal Harvest: The Tragedy of Industrial Agriculture*. Sausalito, CA: Foundation for Deep Ecology/Island Press, 221–26.

Watts, M. (2004) 'Are hogs like chickens? Enclosure and mechanization in two "white meat" filières', in A. Hughes, S. Reimer (eds) *Geographies of Commodity Chains*. London: Routledge, 39–62.

Weiss, T. (2007) *The Global Food Economy: The Battle for the Future of Farming*. London: Zed Books.

Whatmore, S. (1995) 'From farming to agribusiness: the global agri-food system', in R. Johnston, P. Taylor, M. Watts (eds) *Geographies of Global Change: Remapping the World in the Late Twentieth Century*. Oxford: Blackwell, 36–49.

Whatmore, S., Thorne, L. (1997) 'Nourishing networks: alternative geographies of food', in D. Goodman, M. Watts (eds) *Globalising Food: Agrarian Questions and Global Restructuring*. London: Routledge, 287–304.

Wheatley, A. (2000) 'Food and wastewater', in J. Dalzell (ed.) *Food Industry and the Environment in the European Union: Practical Issues and Cost Implications*. Gaithersburg, MD: Aspen, 111–229.

Wilkinson, J. (2002) 'The final foods industry and the changing face of the global agro-food system', *Sociologia Ruralis* 42, 4: 329–45.

Williams, G., Meth, P., Willis, K. (2009) *Geographies of Developing Areas: The Global South in a Changing World*. Abingdon, UK: Routledge.

Windfuhr, M., Jonsén, J. (2005) *Food Sovereignty: Towards Democracy in Localized Food Systems*. Bourton-on-Dunsmore, UK: ITDG Publishing. www.ukabc.org/foodsovpaper.htm

Winne, M. (n.d.) 'Community Food Security: Promoting Food Security and Building Healthy Food Systems', Portland, OR: Community Food Security Coalition, www.foodsecurity.org/PerspectivesOnCFS.pdf

Winson, A. (2004) 'Bringing political economy into the debate on the obesity epidemic', *Agriculture and Human Values* 21: 299–312.

Index

adaptation 232
agri-commodities 7, 24, 314
agricultural commodity prices 22, 24
agricultural modernisation 35–7
agricultural revolution 33
agricultural sustainability 257, 260–1
agricultural treadmill 37, 39
agro-ecosystem 68, 314
agro-forestry 86
air freight 196–8
alternative food networks 272–4, 314
Amazonia 73, 144
animal traction 32–5
animal welfare 264, 269
appropriationism 48, 314
aquaculture 8, 88–91, 108
Aral Sea 124
Asian countries 59–60, 78, 85, 90, 121

beef 148; virtual water content of 127
biodiversity 71; and food supply 100; loss of 100–2
biofuels 10, 138–41, 224–6
bottles 180–1
Brazil 25, 73; ethanol industry 139–40, 238
bread, Chorleywood process 165
British diet 9, 160

Campbell, C. 134
cans, environmental impacts of 177
capture fisheries 5, 8, 89: energy efficiency of 107
car-based shopping 198–9
carbon dioxide (CO2) 116–20, 195, 198
carbon sequestration 119–20
Carson, R. 47
cereals 74–5, 126–8
chemical fertilisers 45–6, 94, 225–7, 259–60

chilled foods 174, 196
China 121; and meat consumption 143
chlorofluorocarbons (CFCs) 172–3
climate change 3, 10; and food production 113; effects 115–16; mitigation options 119–20; and supermarkets 198–9; and food security 228–32
coffee 234–5
Community Food Security Coalition 217, 286
comparative advantage, principle of 22
confined animal feeding operations (CAFOs) see factory farming; see also livestock production
consumer durables and food consumption behaviour 160–1
consumerism 262, 314
consumers 6
contract production 49–50; and waste 201–2
cool chains see chilled foods
corn see maize
cost-price squeeze 37
craftsmanship 274

dam-building 123–4
DDT 46
dietary change 142–3, 220, 222, 239–42
distanciation 263, 314
domestic labour 160–1

ecosystem services, definition 69
embedded water see virtual water
energy analysis (input-output ratio) 105–8
energy inputs to food production 168
energy markets 222–4
entitlements to food 3, 212, 315
ethanol 139–40, 224

ethical food 264–5, 268
Europe and European Union (EU) 54, 113, 140, 150–1,167, 183–4
export production 22, 24–5

factory farming 147–9; and waste stream152–3; 269
Fairtrade 265–7
farmers' markets 276–7
fish 7–8; see aquaculture; see also capture fisheries
food and drink manufacturing 168–9
food and energy 4, 223–6
food chemistry 164
food citizenship 7, 284, 286, 296
food miles 197
food planning 279
food poverty 215–17
food price instability 223–5
food processing 162–3, 168
food safety crises 271–2
food security 11; definitions 211, 214, 217, 315; and energy 222–6; and climate change 228–32; and globalisation 233
food service sector 63–4
food sovereignty 12, 241, 243–6, 315
foreign direct investment (FDI) 29
fossil fuels 4, 103, 105–8, 132–4
fractionation 163
freshwater see water
frozen food 174

General Mills 166
genetic diversity 48, 99–102
genetically modified seeds 256
global agri-food system 2–3, 293–4; dynamics of 26–7; types of integration 24–5; scale and structure 27–9; figure 31
globalisation 21–3, 233, 240–1
global warming potential (GWP): 3, 197; of refrigerants 173, 315
greenhouse gases: atmospheric concentrations 114; agricultural emissions 113, 116–18; from UK food chain 113–14
Green Revolution 39–40, 218
gruppi di acquisto solidale (GAS) 286, 288

Haber-Bosch process 46, 225
high-value horticultural exports 25, 196–7, 236–9
high temperatures 228–31
horizontal consolidation 49, 315
household: budgets 3; consumption 203; food waste 204–5
Howard, Sir A. 96
Hubbert, M.K. 134
hungry, numbers of 210, 214, 216, 225

India 123, 129
integrated pest management 258–9
International Assessment of Agricultural Knowledge, Science and Technology for Development (IAASTD) 232, 244, 256–7,
irrigated agriculture 79, 98, 123–4; see also water

Kenya 25, 201, 235–6

land use 71–2
Latin America 19–21, 59–60, 118
liberalisation agenda 22
life-cycle assessment: explanation of 170–2; of consumer products in EU 183–4; of cottage pie and tomato ketchup 185–7
livestock: feeds 77,150–2; grazing 144, 270; production 4, 77, 87–8; population of main categories by world region (table) 146; and waste streams 152–3; see factory farming
livestock products: environmental impacts 183–4
local food 277
low-input farming 36
luxus consumption 206

maize 76, 101; energy inputs in cultivation 107; and ethanol production 139–40; fractionation of 164
Malawi 227
Malthusian 72, 218
market concentration 44, 47; four-firm ratio (CR4) 51–2
meat consumption 4, 142–3, 220, 222; arguments about, 268–70
mechanisation 41–4

methane (CH4) 116–18,
Mexico 242
Millennium Ecosystem Assessment (MA) 69
mitigation (of climate change) 118–20, 231
mixed farming 86

nanotechnology 182
Nestlé 54, 163, 266
nitrogen 45–6, 95
North–South, definition 8; 28–9
nutrient deficiencies 213
nutrition transition 4, 142, 239, *see* dietary change

obesity and overweight 3, 240, 242
Ogallala aquifer 122
oil *see* peak oil; *see also* fossil fuels
ozone layer, protection of 173, 228

packaging: functions of 175–6; materials used in 176–82
peak oil 10, 133–6, 138
pesticides 46–7, 258
pig (hog) production 50, 150
plastics 178, 182
Pollan, M. 148, 164, 166, 269–70
polyethylene terephthalate (PET) 178, 179
population and food supply 218–22
potatoes 75, 101
primary foods 32, 79
productivism (and productivist) 5, 35, 315
public procurement 282–4

quality attributes 7, 274

rainfed agriculture 83
refrigerants 172–3
refrigeration, rise of 159; 172–5
relocalisation 7, 273, 316
retail distribution systems 192–4
retailers, top five 59, 61–2; 187, 189
retailing 54–5; economies of scale 56–7
rice 75–6, 121, 127–8; and greenhouse gas emissions 118; and pests 259
road freight 193–4

Sahel 230
school food 284–5

scientific knowledge 2, 23
seeds 47–8
segmentation 53
Sen, A. 212
shifting cultivation 84
short food supply chains 275
soil fertility 45–6
soils 91, 94–6, 260
solar energy 103
soybean 73, 76–7, 151
spice trade 16–17
structural adjustment programmes 21–2
Sub-Saharan Africa 72, 118, 145–6
substitutionism 52, 163, 316
sugar 24–5; competition 51; consumption in Britain17–18; production 18–19; and ethanol 140
supermarkets: and food waste 201–4
sustainable consumption 12, 262, 316
sustainable food system, principles 5, 316
sustainability, principles of 5, contrasting perspectives on 251–4
system properties 254–5

tar sands 137
technological change 32–5, 53, 163; *see also* mechanisation
television and food 161
tradition 274
Transition network 288–90

urban agriculture 280–2
United Kingdom (UK) 8–9; and agricultural change 35–7; and greenhouse gas emissions 113–14, 191–2; habitat loss 100–1; Fairtrade retail sales 266; public sector spending on food 283; food transportation 190–4
United States 94, 131, 134; ethanol programme 139–40; and factory farming 147–9, 151; and Ogallala aquifer 122

vegetarianism 271
vertical integration 49, 316
Via Campesina 243
Vietnam 235
virtual water 4, 10, 126–30

Walmart 59, 61, 242
waste: management 179, 181; food 11, 199, 206; from livestock 152–3
water: bottled 179–80, 208; extraction/withdrawal for agriculture 4, 98, 121–6; accounting framework 98; nitrate contamination 98; global share by crop 128; use in food and drink manufacturing 169
water footprints 130–2
wheat 75–6
women and domestic appliances 160–1
World Food Summit (Rome 1996) 214, 243
World Trade Organization (WTO) 22, 245